PRODUCING BEEF
PROFITABLY *from*
PASTURE

PRODUCING BEEF
PROFITABLY *from*
PASTURE

David Hamilton

Copyright © 2015 by David Hamilton.

Library of Congress Control Number: 2015901296
ISBN: Hardcover   978-1-5035-0204-8
             Softcover   978-1-5035-0203-1
             eBook   978-1-5035-0202-4

All rights reserved. No part of this book may be reproduced or transmitted in any form or by any means, electronic or mechanical, including photocopying, recording, or by any information storage and retrieval system, without permission in writing from the copyright owner.

Any people depicted in stock imagery provided by Thinkstock are models, and such images are being used for illustrative purposes only.
Certain stock imagery © Thinkstock.

Rev. date: 01/31/2015

To order additional copies of this book, contact:
Xlibris
1-800-455-039
www.Xlibris.com.au
Orders@Xlibris.com.au
698367

CONTENTS

Introduction 7

Chapter 1	Feed Digestion	9
Chapter 2	Pasture Feeding Value	22
Chapter 3	Pasture Management	31
Chapter 4	Grazing Management	45
Chapter 5	Grazing Behaviour	57
Chapter 6	Climatic Effects	61
Chapter 7	Fodder Conservation	66
Chapter 8	Mixed Stocking	73
Chapter 9	Growth and Development	77
Chapter 10	Milk Production	90
Chapter 11	Calving Season	103
Chapter 12	Fertility	113
Chapter 13	Dystocia	132
Chapter 14	Breeding Principles	144
Chapter 15	Animal Selection	152
Chapter 16	Beef Quality	161
Chapter 17	Parasites	174
Chapter 18	Drought Feeding	182

Appendix   Scoring Methods of Beef Cow Condition 189
About The Author 191

INTRODUCTION

It is clear that many farmers could benefit from better knowledge of research information. Therefore, the aim in this book is to provide better access to research information from various sources that is pertinent to managing grazing beef cattle in southern Australia.

In beef production, the cost of production is commonly at least half the gross return, so any improvement in efficiency, such as from a better calving season, can commonly have at least double the percentage effect on profit that it has on the gross return. Thus it can be more rewarding than it may at first seem be attentive to detail.

In conducting experiments, several replicates of each treatment are usually included, and the effect of a treatment is regarded as significant only if it is large enough relative to any random variation in result among the replicates. Moreover, the results for pasture and other feeds are commonly reported according to the content of dry matter (DM) or organic matter (OM). DM values correct for differences in the moisture level from such things as differences in the amount of rain on pasture, and additionally, OM values correct for variations in mineral content from such things as differences in soil contamination. Also, a heifer is regarded as a young female up to the stage it would be expected to have reared its first calf.

Much of the research on beef cattle management in southern Australia was conducted at Canberra in the Australian Capital Territory, Hamilton in western Victoria, and Rutherglen in north-eastern Victoria, and these locations will be mentioned when this may have merit in indicating the environment that applied.

At Canberra it is relatively cool with substantial amounts of summer rain, at Rutherglen there is a medium rainfall and pastures of annual species, and at Hamilton it is relative cool and the pastures are of perennial species.

CHAPTER 1

Feed Digestion

The value of a feed for cattle depends on the nutrients it contains, how well it can be digested, and how well the animals can absorb and utilize the resulting digestion products.

FEED FRACTIONS

For their nutrition, cattle require energy, protein, minerals, and vitamins. Energy is mainly provided by carbohydrates, such as fibre and sugars present in plants, and is required for such things as building tissues, maintenance of body temperature, fuelling of movement, and other biological processes within the body. Protein is required for building tissues, such as the muscle, skin, and hair, and any protein surplus to requirement is used as energy. Minerals are necessary for various biological processes to occur within the body, and the bones are largely formed from minerals. Vitamins are catalysts which enable various chemical processes to occur within the body. Of the vitamins required, vitamin A is present in green feed and the other vitamins are produced within the animal's body from microbial activity in the rumen or from the action of sunlight on the skin.

The feed requirement of cattle is usually evaluated as a maintenance component and a production component. The maintenance component is required to maintain weight and support the bodily functions. It generally increases by about 73% for each progressive doubling in live weight. It can also vary to some degree with the body condition of the animals and how the weather conditions affect the amount of energy required to maintain body temperature. Some results have indicated that the maintenance requirement at the same live weight may be greater by about 25% for dairy-breed cattle and by about 7 to 12% for beef x dairy breed cattle than for beef-breed cattle, because of the differences in the thickness of the muscles and subcutaneous fat cover and in the size of the digestive tract affecting the levels of heat production and loss.

The nutrient requirement varies with the nature of the production. Young, growing animals have a relatively high requirement for protein, older fattening animals have a relatively high requirement for energy, and cows producing milk have a high requirement for various nutrients.

Feed Digestion

Cattle, like sheep and some other species of animals, have a ruminant digestive system. A notable feature of this system is a four-chamber stomach in which the first two chambers open widely into each other to form a large fermenting vat, technically called the reticulo-rumen but commonly just called the rumen. This system is well adapted to the digestion of fibrous feed, but because of the number of chemical processes involved it is a less-efficient system in digesting concentrated feed than the more direct system found in animals such as pigs.

Feed consumed by cattle is mixed with saliva in the mouth and passes down the oesophagus, or gullet, and into the rumen. In the rumen, the feed particles (called the digesta) are constantly being churned in rumen fluid, and for certain periods of the day boluses of feed are regurgitated for further chewing, a process called rumination,

In the rumen, single-celled microbes consisting mainly or entirely of bacteria and protozoa, attach to the feed particles and excrete enzymes which begin the digestion process.

Much of the protein is broken down to ammonia and some is also broken down to amino acids. The resulting ammonia is used by the rumen microbes to form microbial protein, which later becomes available to the animals as the microbes pass out of the rumen and are digested in the small intestine. Any ammonia not captured by the rumen microbes is absorbed through the rumen wall, converted into urea in the liver, and mostly excreted in the urine. The degree of protein breakdown in the rumen can be proportional to the solubility of the protein.

The energy components digested in the rumen are converted into volatile fatty acids consisting mainly of acetic, propionic, and butyric acids. The proportion of acetic acid generally increases with the digestion of more fibrous feed, and that of propionic acid increases with the digestion of more-digestible feeds, especially grain. The relative proportions of the different volatile fatty acids affect how efficiently the feed is used for different purposes. A relatively high level of propionic acid can result in lower milk production and lower fat content in the milk of cows, and it favours the deposition of body fat.

Fat, which is usually only a minute component of pasture, is not digested in the rumen but is converted from the unsaturated form found in plants into the saturated form that is less desirable for human nutrition when consumed as animal products.

Large volumes of carbon dioxide and methane gases are also produced during digestion in the rumen, and it has been estimated that 7–12% of the feed energy can be converted into these gases. These gases normally pass through the oesophagus and are exhaled, their passage being controlled by sensors located at the base of the oesophagus.

The feed particles remain in the rumen until they are reduced to about the size of tea leaves. More-digestible feed is digested more quickly and passes more quickly out of the rumen. With fibrous feed, grinding or chopping can result in a moderate increase in the speed of digestion and in the amount of feed consumed, but a slight decrease in the degree of digestion. The benefit from grinding can be greater with younger animals because of their less-developed digestive systems.

Typically, when cattle are consuming feed such as immature pasture, 65–70% of the energy digestion occurs in the rumen, 20–30% in the small intestine, and 5–15% in the caecum and colon. With less-digestible feed, such as mature pasture, a greater proportion of the digestion occurs in the rumen, and more of the digested energy is used in the digestion process.

In sequence in the next three stages after the rumen, water is removed, the remaining material is acidified to aid protein digestion, and bile from the bile duct in the liver is added to neutralize the acidity and aid fat digestion.

Following this is a long narrow tube called the small intestine, where enzymes secreted by the gut wall act on the remaining feed particles and the digestion products are absorbed through the gut wall. The digestion process here is more direct and more efficient than in the rumen. The main digestion products absorbed in this part of the digestive system are glucose, amino acids (the building blocks of proteins), and the products of any fat digestion.

Finally, there is the colon with the caecum attached. Here there is further digestion by microbes and absorption of volatile fatty acids, but the resulting microbes are passed out in the dung, so their nutritional value is lost.

Because of the amount of protein that can be lost in the rumen and the amount converted into microbial protein, the protein that is finally digested and absorbed in the small intestine may be deficient in both quantity and composition for maximum animal performance. If this occurs, animal performance may be improved by providing a feed in which more of the protein is not readily digested in the rumen but is digested in the small intestine. This is called bypass protein. The proportion of bypass protein varies between feeds, and is generally relatively high in animal byproducts, such as meatmeal and fishmeal. The degree of protein bypass can also be increased by heating the feed or treating it with chemicals. The animals most likely to benefit from this protein are those with a relatively high protein requirement, such as young growing animals and cows producing milk.

Proteins are rich in nitrogen and the nitrogen level differs a little between them. Accordingly, the crude protein content of a feed is assessed as 6.25 times the nitrogen content and the nitrogen content is sometimes shown rather than the crude protein content.

When the feed is low in protein the rumen microbes can use nitrogenous compounds present in the saliva or entering the rumen from the bloodstream to form microbial protein, and thereby partly compensate for the low protein consumption. In the same way, the rumen microbes can also use non-protein nitrogen, notably urea and ammonia added to the feed and non-protein nitrogen already in the feed. The use of non-protein nitrogen by the microbes depends on sufficiency of energy and minerals present to support their activity, with sulphur being an especially important mineral in this regard.

Urea added to the feed can pass through the rumen within about half an hour. Consequently, it can be used more effectively if dissolved in water and the solution is allowed to soak into the feed. Urea in large amounts is toxic, so it should be thoroughly mixed with any feed it is added to.

Biuret can be used instead of urea. This product is usually retained longer in the rumen, but an adaption period for the animals is required to use it effectively.

Bloat

When cattle are consuming some legumes, the gas produced in the rumen can become entrapped in bubbles which fail to activate the sensors at the base of the oesophagus that control the release of gas. This can result in a greatly distended rumen, causing the condition known as bloat.

The tendency to cause bloat can vary among legumes and between the same legumes at different times of the year, because of differences in the chemical composition of the plants as a result of weather changes or plant maturity. The amount of surface water on the plant can also affect the likelihood of bloat, and it has been claimed that lucerne should not be grazed by cattle until the dew has lifted in the morning.

Some cattle bloat more readily than others and this seems to be under genetic influence. The cattle can experience a persistent moderate level of bloating or sudden acute bloating. Persistent bloating can reduce pasture consumption, and vets have claimed that the cattle experiencing the acute form are the ones most likely to die.

Dosing the animals with anti-bloating agents, such as tallow and the detergent Teric 12A23, releases the gas. In critical cases the rumen can be pierced with a cannula and trochanter to release the gas, or else a soft hose about 12 mm in diameter can be passed gently down the gullet and withdrawn and reinserted if resistance to its passage is felt.

There are various ways of administering anti-bloating agents, including the use of slow-release capsules inserted into the rumen. Providing hay can be an ineffective treatment unless the pasture conditions are such that cattle will eat much hay or an anti-bloating substance such as tallow is added.

Intake

The factors affecting the animal's level of consumption are complex and not fully understood. Generally the consumption of fibrous feed, such as pasture, is limited by the speed of digestion, and the consumption of concentrated feed such as grain is limited by the animal's appetite for nutrients. Also, any deficiency in the diet, such as a deficiency of protein or a mineral, can affect the level of feed consumption by affecting either or both the actions of the rumen microbes and the metabolism of feed within the animal's body. Moreover, very fat animals and pregnant animals can have comparatively low levels of consumption. Possible causes suggested for these effects have been: a) increased pressure on the digestive tract, b) an inherent mechanism which restricts consumption once a certain body condition has been reached (as in the case of fat animals), and c) a change in hormone levels that reduces appetite (as in the case of pregnant animals). The occurrence of greater metabolic activity within the animal can also result in increased feed consumption as was evident in one instance when lactating cows consumed 47% more feed than similar dry cows. The level of consumption can also be affected by unpalatable factors in the feed, the high acidity levels of some silages, and the high rumen acidity levels with large grain rations. As part of the complexity, the peak in pasture consumption by dairy cattle can be later in the lactation than the peak in milk production.

GRAIN DIGESTION

Grain is sometimes fed to cattle as a pasture supplement. It is also used in intensive feeding and in drought rations.

The nutritional value of grain differs among grains and among different samples of the same grain. Better growing conditions usually result in plumper grain and greater starch content, and the crude protein level can be proportional to the nitrogen status of the soil on which the grain was grown.

Some of the reported values from the composition of oats, barley, and wheat dry matter are, respectively, 8.3–19.1%, 7.8–16%, and 7.1–22.8% for the crude protein content, an average of 63%, 76%, and 81% for the starch content, and 10%, 5%, and 2% for the average crude fibre content. Lupins are sometimes also fed to beef cattle and contain about 30% crude protein.

Grinding or rolling the grain can result in an improvement in digestibility of about 7% for oats, 23% for barley, and 32% for wheat. In the case of maize and sorghum, the digestibility can be further enhanced by heat treatment to pop the grain or by using high-pressure rollers to micronize the grain. In one study the improvement in animal performance from steam-flaking of maize rather than dry-rolling was 6%. Also, the starch in barley and sorghum can be gelatinized by treating the grain in water at a temperature of at least 58 to 64 °C for barley and 67 to 77 °C for sorghum. Moreover, some dark coloured varieties of sorghum can contain toxic levels of tannins.

Cattle given free access to grain can gain weight just as rapidly on whole grain as on processed grain, but they will eat more grain for the same weight gain. When heifers at Canberra were fed a mixture of 20% lucerne chaff and 80% whole or crushed wheat the amount of feed required for the same weight gain was 24% greater with the whole grain, and at the same level of feeding the weight gain was 23% lower with the whole grain.

Young calves can generally digest whole grain better than older cattle, and the degree of digestion of whole grain may vary with the proportion of grain in the diet. In one instance, the proportion of the whole maize ending up in the dung was 18% when the ration contained 14% of hay and 28% when the ration contained 28% of hay.

The different rumen microbes are specific in the digestive function they perform, so the rumen population of microbes needs time to adapt to a marked change in the composition of the diet, especially if grain is being added. If the change in diet occurs too quickly there can be adverse effects ranging from a small reduction in performance to more serious effects.

Blood tests have shown that it may take up to about six weeks for the rumen population to fully adapt to grain digestion and up to about two weeks to fully adapt to a change in forage diet, but a substantial degree of adaptation can be achieved before the end of these periods.

If grain is introduced too quickly into the ration toxic levels of lactic acid can form in the rumen. Later, high levels of grain can result in acidity levels that damage the rumen wall and enable bacteria to pass through and form abscesses in the liver. Other effects can be lameness resulting from the production of histamines damaging the blood vessels of the feet as well as hair consumed by licking becoming embedded in the inflamed rumen wall, and the pH level of the blood reaching a fatal level.

Anything that speeds the digestion of the grain, such as finer grinding, also increases the risk of acidosis. For feeding cattle the grain should only be cracked. Finer grinding may also reduce the amount of rumination and worsen the overall problem. In a British experiment, calves being fed ground and pelleted dried grass could not be kept healthy until chopped dried grass was included in the diet to start rumination.

Inclusion of hay along with grain in the diet reduces the likelihood of acidosis by reducing the amount of grain consumed and by stimulating rumination. Also, chewing hay, especially if the hay is long rather than chopped, increases the flow of saliva during chewing, and saliva contains bicarbonates and phosphates which help to moderate the rumen acidity level. An increase in the crude protein level of the diet can also reduce the likelihood of acidosis, because the ammonia produced during digestion helps to reduce the pH level in the rumen and enables the rumen microbes to assimilate the products of digestion more actively.

Adding an ionophore to the grain can reduce the likelihood of acute lactic acid poisoning because of reduced feed consumption and a change in the digestion products. The three commonly used ionophores are monensin, lasalocid, and laidlomycin propionate, with trade names respectively of Rumensin, Bovatec, and Cattlyst. They commonly result in about a 10% reduction in feed consumption with no effect on weight gain. They act by: a) reducing methane production, b) suppressing the activity of less-desirable rumen microbes, c) reducing protein breakdown in the rumen, and d) resulting in a more favourable balance in the volatile fatty acids produced.

The inclusion of sodium bicarbonate in the diet usually has no appreciable effect on the result because the bicarbonate has usually left the rumen before the peak in acidity level is reached about six to ten hours after feeding.

Different grains and different samples of the same grain can differ in the chemical composition of their starch, and this affects the speed of digestion and the likelihood of acidosis. Because of this effect, wheat and grains used in brewing or distilling can be relatively dangerous to use. Also, with maize and sorghum the starch is surrounded by protein if the kernel is of the flint form rather than the dent form, and this affects the proportion of starch digested in the rumen. In one experiment, when steers were fed hay and cracked wheat, dent maize and flint maize, the respective values were 86.6, 60.8, and 34.8% for the apparent digestion of starch in the rumen and 5.16, 5.81, and 5.94 for the minimum pH level in the rumen.

Cattle can be fed lupins at up to about 85% of the diet along with hay without conditioning, but with a greater proportion of lupins than this there can be grain compaction in the rumen from the grain swelling.

In feedlotting, cattle are commonly conditioned and then given continuous access to a complete mixed feed to even out consumption and the level of rumen acidity.

Stress on cattle from such things as cold, wet weather and temporarily running out of water can also lead to acidosis, and any change between grains should be done gradually.

Mineral Nutrition

Mineral nutrition is an important part of the overall nutrition of the animals, and periodically it attracts considerable attention for various reasons.

Minerals are required for the formation of bones and tissues and the operation of various metabolic processes within the animal's body. Minerals are also important for the functioning of the rumen microbes, so any imbalance can affect the animal directly or indirectly.

With grazing animals the mineral content of the pasture can vary with numerous factors, including the soil type, soil pH, fertilizer application, plant species, degree of plant maturity, and seasonal weather conditions. The degree of soil contamination of the pasture and the mineral content of the drinking water can also affect the result. In north-eastern Victoria an iodine deficiency in young lambs mainly occurs during spring, because during winter the lambs generally receive adequate amounts of iodine from the soil ingested when grazing short pasture. The amount of soil consumed can also affect the cobalt status of the animals.

The animal's mineral requirement varies according to the nature and level of its production. For example, young growing animals have a higher mineral requirement than more mature animals, and lactating cows have an especially high mineral requirement. Moreover, animals just maintaining weight on dry pasture over summer have a lower mineral requirement than those gaining weight rapidly on green pasture during spring. Worm infestation can also affect the animal's mineral status.

The mineral content of the herbage differs substantially between plant species, and as plants mature there is a general decline in their mineral content. Marked differences in the botanical composition of the pasture between years can affect the mineral nutrition of the grazing animals. For example, in north-eastern Victoria grass tetany in beef cows, which is primarily due to magnesium deficiency, is mainly a problem in years when late opening rains result in a low clover content in the pasture or the cows are grazing growing oats. Another example is that in the Murray River basin when clover seeds germinate relatively early in the year the seedlings growing during the comparatively hot weather can absorb levels of copper that are toxic to sheep.

Aside from the level of a particular mineral element present in the plant it is also important how well it can be absorbed and utilized by the animal. The degree of absorption and utilization can vary greatly between mineral elements and can be affected by the stage of growth of the plant and the levels of other elements in the diet. For example, the absorption and utilization of magnesium is affected by the levels of sodium, phosphorous, and potassium in the herbage. Also, the metabolism of copper in the animal is affected by the levels of sulphur and molybdenum in the pasture, and in one experiment, the application of lime to the soil resulted in an increase in the level of molybdenum in the pasture that induced copper deficiency in sheep.

Some minerals can compensate to some degree for a deficiency of other elements. For example, calcium can partially substitute for a deficiency of magnesium and potassium for a deficiency of sodium.

A deficiency of calcium, causing milk fever, is a common problem with cows of high milking capacity. The problem is not directly due to a deficiency of calcium in the diet but instead is due to inadequate mobilization of calcium reserves from the bones. The feeding of calcium before calving can actually dampen down the mobilization mechanism and worsen the problem. Research has been directed at improving the calcium-release mechanism.

Some plant toxins can also affect the mineral nutrition of the animals, with a notable example being the liver damage and consequent copper toxicity in the blood that can occur in sheep consuming the summer-growing weed heliotrope. The provision of licks containing molybdenum can help to reduce the copper levels in the blood.

Animals have substantial body reserves of minerals and an effect on their performance generally does not become apparent until the reserve of a particular mineral element has been depleted. Also, as already indicated, in any one location a particular mineral problem may be persistent or may only occur occasionally under specific circumstances.

If a mineral problem is suspected it is vital to consult someone with adequate knowledge to understand the possible complexities involved. Chemical tests on the soil can indicate the likelihood of a possible deficiency of a mineral such as cobalt but these tests do not provide accurate assessment of the likely mineral content of the plants. Also, plant analysis can indicate possible deficiencies, but sometimes the mineral content of the plants does not closely indicate what is being absorbed and utilized by the animal.

A more detailed study requires animal tests, but intensive testing can be prohibitively expensive because of the variation in the results between animals, the period over which the tests have to be conducted, the different kinds of samples required for different tests, and the variation in the results that can occur with season and small changes in the pasture conditions. It is also important to recognize that specific mineral problems require specific treatments, and that indiscriminate provision of mineral supplements may be harmful or wasteful.

There are now well-defined regions in Australia where particular mineral problems may sometimes be encountered. If there is a serious mineral problem, this generally becomes apparent at some stage.

Conclusions

The ruminant digestive system of cattle is well adapted to the digestion of fibrous feeds, and care is required in using more concentrated feeds.

The nutritive value of feed depends on the speed and degree of digestion and the nutrients provided relative to the nutrient requirements of the animal. Aside from meeting the nutritional needs of the animal, it is also important to meet the nutritional needs of the rumen population of microbes involved in feed digestion.

The rumen's population of microbes needs time to adapt to a marked change in the chemical composition of the feed, especially if grain is to be provided. Grains differ in their degree of safety, and increased processing can increase the risk. The inclusion of hay or other roughage and a digestion modifier, such as monensin, can be beneficial.

Bloat in cattle grazing legumes can be due to non-activation of censors which control the release of gas.

Mineral problems sometimes occur in grazing animals. Regions where such problems are most likely to occur are now well documented. Advice should only be sought from adequately trained people.

CHAPTER 2

Pasture Feeding Value

The quantity and quality of the pasture present affect the performance of grazing cattle.

SEASONAL EFFECTS

Results obtained at Rutherglen for steers kept on annual pasture at several stocking rates for a year after weaning in early summer show how different pasture conditions at different times of the year can affect cattle performance. The results were as follows:

a) generally an increase in weight of about 5 and 10 kg on dry pasture from early summer until about the end of February
b) maintenance of weight or a loss in weight for the remainder of the period of dry pasture, depending on how much pasture was present
c) a marked improvement in animal performance with the start of a green pick about two weeks after the opening rains
d) for the remainder of the period until the end of winter a range from a weight loss to a rapid weight gain, depending on the amount of green pasture present
e) rapid weight gain during spring, with the steers at the heavier stocking rates showing compensatory growth.

In years with a late start to pasture growth the steers at the lighter stocking rate were still generally able to maintain weight on dry pasture until late autumn, even though by then the pasture had developed a darkened, weathered appearance. Also, any green pick during summer and early autumn as a result of unseasonable rainfall usually resulted in improved animal performance for the period as a whole.

The same pattern of response relative to the pasture conditions has generally applied in other regions, except that in some instances the animal performance on dry pasture appears to have been poorer.

In the case of cows, pregnancy and any milk production have additional effects on the weight change, and at Rutherglen Angus cows calving at the start of summer lost weight at the rate of a kilogram a day or more on the plentiful dry pasture which followed.

A misleading effect with young steers grazing dry pasture is that because of continuing growth in frame size the steers lose body condition and appear to lose weight even when maintaining weight.

Green Pasture

Winter pasture has relatively high moisture content and relatively low sugar content, but it can be digested rapidly and this benefits intake.

Results with sheep fed unlimited pasture in pens have indicated that the moisture content of winter pasture may not greatly restrict the nutrition of grazing animals, except perhaps in the case of plants with especially high moisture content such as capeweed. Thus the main limitation to animal performance during winter is generally the quantity of pasture present. However with animals in pens there has usually been slightly better performance on spring pasture than on winter pasture because of a higher level of soluble sugar in spring pasture, but with grazing animal factors other than the quality of the pasture can influence the result.

Steer growth at Rutherglen relative to pasture height was generally as good during winter as during spring.

Late in the growing season, generally near the time of flowering, pasture plants start to mature. This is accompanied by an increase in the fibre content of the plants and a decrease in the mineral and protein contents. The actual timing of these changes can differ to some degree between plant species, with the clover sometimes starting to mature a little later than the grasses, but the eventual decline in digestibility can be about the same for both. The effect on the animals is generally a reduced degree of digestion, slower digestion, reduced intake, a greater amount of energy expended in eating and digesting the pasture, and less nutrients obtained from each kilogram of feed consumed.

Some reported changes in the properties of several annual pasture plants at different stages of maturity were as follows:

	Clover	Capeweed	Ryegrass	Other Grasses
Moisture (%)				
20 August	81.5	92.8	86.3	83.7
Flowering	85.7	92.8	86.3	83.7
Digestibility of dry matter (%)				
20 August	61.3	67.4	61.3	60.4
Flowering	60.9	64.2	59.2	54.7
Dry	35.8	51.1	39.6	36.9
Crude protein in dry matter (%)				
Flowering	19	17.2	12.5	14.5
Maturity	12.1	12.5	8	10

Aside from affecting the nutritional value of the standing pasture, these changes also greatly affect the quality of any conserved fodder made from the pasture.

In one experiment in which sheep in pens were fed *Phalaris* at advancing stages of maturity, the results were a decline from 1,067 to 804 g a day for the consumption of organic matter, an increase from 4.11 to 5.78 hours a day in the time spent eating, and an increase from 9.5 to 17.5 kcal/100 g organic matter for the energy spent in eating and digesting the feed.

When pasture is being grazed continuously cattle can largely counter the effect of increasing plant maturity during spring by selectively grazing the more nutritious sections of the pasture. This selective grazing is often seen as the cattle concentrate their grazing on patches of the pasture, with this heavier grazing slowing plant maturity on these patches. However, as the pasture dries off the opportunity for selective grazing declines sharply, as does animal performance.

Lambs on green pasture usually grow faster on clover than on grass. In one experiment when sheep were fed eight grasses and six legumes the level of consumption was 28% greater on average with the legumes than with the grasses. The superior result with the legumes was attributed to tighter packing of the herbage in the rumen and faster digestion. The continuing response of lambs to relatively high protein levels may also have contributed to the result. However, in experiments with cattle the effect of having clover in the pasture has ranged from beneficial to harmful. In America, the average weaning weight of calves was 161 kg for those grazing mixed clover/and grass pasture and 146 kg for those grazing grass pasture fertilized with nitrogenous fertilizer. On the other hand, at Rutherglen, the botanical composition of the pasture did not significantly affect steer growth relative to pasture height, and in results from elsewhere cattle performance has sometimes been poorer on pasture containing clover. It has been suggested that this last result probably applied because of persistent mild bloat from the clover restricting pasture consumption.

Dry Pasture

On dry pasture there is still some opportunity for selective grazing, but this declines as the more nutritious fractions are eaten.

The crude protein level of dry pasture is generally in the range of about 5.5 to 13% in proportion to the clover content. A protein level below 7 or 8% can result in slower feed digestion and reduced intake. However at

Rutherglen the differences in the protein content of dry pasture between years did not greatly affect the weight change of weaner steers. On the other hand at one location when sheep in pens were fed dry-pasture fractions the daily weight losses were –163 g on ryegrass, –98 g on clover tops, and –33 g on clover burr. Cattle can eat little clover burr even when it is thick on the ground, unless there is little other pasture.

Animals in pens fed dry, mature grass have generally shown increased consumption from being given a protein supplement, but when grazing sheep were given this kind of supplement there was a reduced incentive to graze. Providing licks based on urea have generally been of little or no benefit, as is described in another chapter.

Rain falling on dry pasture can have little effect on the nutritional value of the pasture unless moist conditions last long enough to cause rotting. Clover-dominant pasture is especially prone to rotting because it packs down readily when wet and the high protein content also favours rotting. In one instance, a number of rainfall events of 5 to 35 mm over three days reduced the digestibility of the pasture by 6%.

On perennial pasture at Hamilton the regrowth after haymaking has been more nutritious over summer than the other pasture.

Sheep grazing pasture on poor soil in Western Australia have sometimes benefited from being fed mineral mixtures during summer and the fertility of cattle on tropical pastures has been improved with a phosphate supplement. However, similar benefits do not seem to have been reported with fertilized pasture on the better soils of south- eastern Australia.

Pasture Height

In Britain, steer growth started to decrease when the pasture height dropped below 6 to 8 cm on dense, continuously grazed pasture, and below 8 to 10 cm on the more open pasture occurring with rotational grazing. Also, steer growth in Hamilton was faster on pasture 6 to 16 cm tall than on shorter pasture, and on tropical pasture it has been recorded that a more open structure in the pasture can restrict pasture consumption.

The growth of young steers kept at several stocking rates at Rutherglen was assessed against the median (middle) reading for pasture height, a value that excludes pasture being avoided around dung pats. The steers were generally able to maintain weight on pasture about 20 mm. high. Later in the year the steers generally achieved a weight gain of 1.2 kg a day on pasture of about 40 to 60 mm high but those at the heavier stocking rate in leaner condition continued to respond to taller pasture, and reached a weight gain of 1.6 kg a day. This result for these last steers was probably mainly due to the greater opportunity for selective grazing that accompanied the taller pasture in the situation where the steers had the growth capacity to respond.

These last results were obtained on continuously grazed pasture. If pasture is allowed to become rank before cattle have access to it then animal performance is likely to be reduced because of the overall maturity of the pasture.

Toxic Properties

The following common pasture plants sometimes develop toxic properties:

Capeweed. Very occasionally in autumn these plants can accumulate highly toxic levels of nitrite if a dry spell follows a period of rapid growth.

Phalaris. A period of rapid growth after a dry spell can result in toxic properties in this grass, causing the condition known as *Phalaris* staggers. This can take two forms: one in which the animals when mustered may collapse and die suddenly or recover quickly, or the second form in which a more prolonged period of uncoordinated movement can terminate in death. The problem is due to the accumulation of high levels of nitrogen in the plant together with a toxin which inhibits the animal's metabolism of ammonia derived from rumen digestion. The concentration of ammonia in the blood can cause brain damage. The latter condition is most common in animals grazing pastures on soils low in cobalt, and administering cobalt bullets to the animals can be beneficial in protecting against chronic *Phalaris* toxicity but not against acute toxicity. Newer varieties of *Phalaris* are less toxic than the older varieties.

Perennial ryegrass. The presence of a fungus in the plants can cause uncoordinated movement called perennial ryegrass staggers. Affected animals need to be protected against thirst and starvation and physical hazards such as dams. The cause is the presence of a toxin mainly located at the base of the leaf sheaths. Heavier stocking resulting in closer grazing can increase the problem.

The concentration of the fungus in the seed decreases progressively, and it can be low in seed stored for about 15 months before sowing and absent in seed held for 2 years. Treatment of the seed with a fungicide can also be beneficial. Ryegrass with low levels of fungus or free of fungus is becoming increasingly available.

Another disease occasionally found in animals grazing on pastures containing perennial ryegrass and sometimes with other plant species is scabby mouth. This disease mainly affects sheep, but it can also affect cattle. It is most common in late summer and autumn under moist conditions. The death rate can be high in sheep, but it is usually low in cattle. The disease is caused by spores of a fungus which grows in moist litter. A toxin in the spores damages the liver, leading to photosensitization, which leads to other conditions, such as scabby mouth.

Paspalum. A problem can arise because of the seed heads becoming infected with a fungus. This can cause nervous spasms and affect coordination, but the result is seldom fatal.

Annual ryegrass toxicity. This highly toxic effect leading to death is found in Wimmera ryegrass in regions of Western Australia, South Australia, and Victoria. The problem is due to a toxin produced by bacteria which infect nematodes located in the seed heads of the plants. The toxin may persist for 200 days or more in silage but decreases faster in hay. The problem is likely to occur more in sheep than in cattle because of the greater consumption of ryegrass heads by sheep.

Bracken. Bracken contains a substance which causes thiamine deficiency in grazing animals and may cause tumours and haemorrhage. The plant is eaten mainly in the young frond stage and if other feed is scarce.

Heliotrope. This annual summer weed can cause serious liver damage, especially in sheep. This results in toxic levels of copper in the bloodstream. The effect of eating this weed may not appear until two to three years later. Providing mineral blocks containing sulphur and molybdenum can help to remove excess copper from the blood.

Paterson's curse (*Echium plantagineum*) can have a similar but less severe effect as heliotrope. Horses are reputed to be especially sensitive to the effects of this weed.

Lupins. The stubble of lupins can often be highly toxic because of infection with a fungus. The seed can also be toxic, but generally a substantial amount of lupin seed can be consumed without harm. Cattle are usually less affected by this problem than sheep.

CROP RESIDUES

In cereal stubbles the straw itself is of generally low nutritional value, and the main value of stubbles is in the shed grain and weed fractions present. Sheep are generally better at selecting these items than cattle.

The feeding value of straw can vary greatly. In one study the respective ranges in the crude protein values were 1.5 to 4.9% for wheat straw, 0.9 to 4.7% for oat straw, and 2.5 to 4% for barley straw where the straw hadn't been weathered. Where it had been weathered the values were 2.3 to 3% for wheat straw and 2.5 to 2.8% for oat straw. Overseas results for the digestibility of the dry matter were 36 to 49% for wheat straw, 45 to 55% for oat straw, and 43 to 50% for barley straw. In one set of Australian tests pea straw had a higher crude protein level than oat straw but the digestibility was similar for both.

In one experiment, when Hereford heifers grazing wheat stubble were given a supplement in roller drums of 27% molasses, 5.1% urea, and 0.23% sulphur and free access to a mineral lick, the daily weight loss was reduced from −0.26 to −0.11 kg. It was estimated that the improved performance with the provision of the lick was entirely due to the amount of molasses consumed and not to any marked improvement in straw consumption or digestion.

When yearling steers in South Australia were given access to crops that hadn't been harvested their weight gain remained good only as long as substantial amounts of grain remained in the heads of the plants. The average daily weight gains over 100 days were 0.98 kg on field beans, 0.55 kg on lupins, 0.23 kg on Cyprus vetch, 0.25 kg on field peas, and 0.16 kg on subterranean clover. Lambs treated similarly gained weight at about 0.2 to 0.25 kg a day on all except the subterranean clover, where the weight gain was 0.06 kg a day.

In another experiment, steers grazing wheat or barley crops over summer had daily weight gains of 0.71 and 0.96 kg a day respectively over 51 to 86 days. It was estimated that they consumed 64 to 80% of the grain.

Conclusions

The value of pasture for grazing animals is mainly affected by the quantity present and its nutritional value, which depends greatly on the degree of plant maturity. The main effect of rain falling on dry pasture is rotting from prolonged wetting, especially on clover-dominant pasture. Some pasture plants can sometimes develop periods of toxicity.

Steers may grow well for a period on crops that haven't been harvested but cereal stubbles can be of little value for cattle.

CHAPTER 3

Pasture Management

Several farm assessments have shown that the profitability of a grazing enterprise is usually proportional to the productivity of the pastures and the effect this has on the carrying capacity.

SPECIES SELECTION

Perennial species are sown in regions of high rainfall and annual species in regions of lower rainfall. Also, a perennial grass can sometimes be sown in regions of medium rainfall if the soil conditions are suitable.

The main perennial grasses sown are perennial ryegrass in regions of high rainfall and *Phalaris* in regions of lower rainfall. Cocksfoot and tall fescue are perennial grasses that are also sometimes used. *Phalaris* is usually more persistent than perennial ryegrass in marginal regions and fescue can respond relatively well to summer rain.

The annual grass Wimmera ryegrass used to be widely sown but is now seldom sown, because of possible toxicity and difficulty in controlling it in crops sown in a rotation with pasture.

The legumes sown are white clover in regions of high rainfall, subterranean clover in regions of lower rainfall with acid soil, and medics in regions of low rainfall with alkaline soils.

There are also various annual legumes and annual ryegrasses that are used for temporary pasture, mainly for haymaking.

Important Attributes

In seed selection some important plant attributes to consider are the following:

- Aside from total production seasonal production can also be important. Winter is a time of scarce pasture and any increase in pasture growth at this time of year usually affects directly the amount of animal production for the year as a whole. Different varieties of *Phalaris* and subterranean clover are among those that differ in their seasonal growth pattern. There can also be benefits from using plants that continue to respond to late rains, and one of the benefits of Wimmera ryegrass in annual pasture is that it can respond to late rains in years when an earlier dry spell has sent the other pasture plants to seed.
- It is important that the plants persist. *Phalaris* and Wimmera ryegrass can persist better under the lighter stocking that commonly applies with cattle than with sheep, and differences in the seasonal growth pattern of *Phalaris* varieties can also affect the persistence under some grazing methods.
- Some plants are more tolerant of some soil conditions than others, and Trikkala subterranean clover, *Phalaris*, and strawberry clover are relatively tolerant of wet conditions.
- The maturity time is very important with subterranean clover varieties. If the plants mature too early there can be lost production and considerable rotting of the clover leaves before the pasture plants as a whole dry off, and if the maturity is too late the seed set and persistence can be poor. The hardness of the seed of subterranean clover varieties can also greatly affect

persistence by affecting how much viable seed remains after false breaks or years unsuitable for the germination of this seed.
- The resistance to some diseases can differ between different varieties of the same plant species. This is the case with regard to the susceptibility of some older varieties of subterranean clover to some root and leaf diseases.
- In the case of perennial ryegrass and *Phalaris*, the occurrence of toxic properties is less likely with some of the newer varieties.

Improved varieties of pasture plants are being continually developed, and recommendations can be obtained from local agronomists and seed merchants.

Pasture Establishment

Pasture seeds are expensive but using an adequate seeding rate to achieve a fairly dense pasture quickly can result in much greater pasture production during the first two years, and reduced opportunity for weed invasion.

Especially with legumes, the seed can germinate better and the plants establish faster with warmer soil conditions. With autumn sowing this can greatly affect how well the plants can resist the rigours of winter. With subterranean clover seed the level of germination can decline rapidly with delayed sowing after the opening rains and with a later opening to the season.

With legumes it is desirable to inoculate the seed with rhizobium bacteria if the soil has not carried these plants in recent years, and on acidic soil there is likely to be additional benefit from coating the seed with lime and adding the trace element molybdenum to the soil. With very acidic soils there can also be a benefit in using a mixture of lime and superphosphate at sowing.

Pasture seeds are often sown under cereals. This can result in some smothering of the young pasture seedlings as well as the clover seed being sown too late for an optimum result. Some varieties of subterranean clover establish better than others under cereals, and with the common cereals the risk of smothering can be lowest with triticale and highest with oats.

The seedlings of some perennial grasses, particularly *Phalaris* and tall fescue, compete poorly in their early stages, so with these species it is important to use an overall seeding rate that will limit the degree of plant competition during establishment. It is also most important not to sow pasture seeds more than 1 cm deep.

With newly-sown pasture seeds the resulting plants should not be grazed until their roots are firmly established in the soil. It is best to graze at a level that will prevent any grasses from smothering the clover. A recommendation with perennial ryegrass is not to graze before the fourth-leaf stage. With subterranean clover the plants can sometimes set more seed if grazed moderately in early spring, but not later. Sheep grazing heavily on Wimmera ryegrass during seed set can reduce the proportion of this grass in the pasture. With *Phalaris*, allowing the plants to set seed can result in the development of stronger buds at the base of the shoots, leading to larger and more-robust plants in the following year.

Stocking newly-sown pasture with sheep rather than cattle for the first year can sometimes result in smoother paddocks.

Annual v Perennial Grasses

Perennial species have a clear production advantage over annual species in regions of high rainfall but the merit of combining a perennial grass with annual species in regions of medium rainfall is less clear.

The main perennial grass used in regions of medium rainfall is *Phalaris*, but tall fescue or cocksfoot is sometimes used as an alternative.

In medium-rainfall environments at Canberra and Wagga Wagga and in Western Australia, using a perennial grass in sheep production was only beneficial in years of low rainfall. A benefit generally seemed to occur if there was sufficient rain during late summer and autumn to produce green shoots from the perennial grass but insufficient rain to produce germination and growth from the annual species. In Western Australia, continuous heavy grazing by sheep during summer reduced the survival of the perennial pasture plants. On the other hand, in the two experiments in Western Australia the increase in wool production from using a perennial grass was 37% with *Phalaris* and 18% with cocksfoot.

At one stage it was thought that the inclusion of a perennial grass would result in a faster response to the opening rains, but this did not prove to be the case.

Despite these effects there can still be advantages in using a perennial grass in marginal regions in situations where the soil moisture levels are likely to be adequate. These advantages can include (a) lower invasion of volunteer species,(b) better persistence after haymaking, and (c) possible slowing of the acidification of the soil. On the other hand, disadvantages can be: (*a*) more-difficult establishment and a greater establishment cost, (*b*) sometimes poor persistence under continuous heavy grazing, (*c*) the formation of large clumps of unpalatable herbage in spring if the grazing is too light, and (*d*) occasional toxic properties. Because of the greater establishment costs perennial grass should generally be used only in a long-term pasture.

It may be that perennial grass can help to reduce the acidification of the soil by reducing the amount of soluble minerals lost in the drainage water. It could also act by reducing the amount of clover in the pasture.

Phalaris has been reported as having deeper roots than perennial ryegrass, and this may help its persistence.

Fertilizers

Adequate fertilizer application is necessary to achieve highly productive pastures.

The main chemical elements applied in fertilizers are nitrogen, phosphorus, potassium and sulphur, with phosphorus and sulphur being the main ones by far used in grazing enterprises. Periodically trace elements are also often applied.

The most commonly used trace elements are copper, zinc and molybdenum, but others are required on some soils. The likely trace element requirements for many regions have now been well mapped. These elements may benefit either or both plant growth and animal health.

The greatest requirement is for phosphate fertilizer. In the past this fertilizer was almost always used as single superphosphate, containing about 9% phosphorus and about 11% sulphur. Now double and triple superphosphate are also being used, and respectively they contain about 17.5 and 20% phosphorus and about 3.5 and 1.5% sulphur. Legumes have a relatively high requirement for sulphur, and when these last two fertilizers are used additional sulphur can be added to the soil by coating the phosphate fertilizer with different levels of raw sulphur or by applying gypsum. Raw sulphur is slow-acting during cold weather and has an acidifying effect on the soil that requires about 3kg or lime to neutralize each kilogram of sulphur applied.

Different forms of ground phosphate rock containing about 11 to 16% phosphorous are also available, but they are usually extremely slow-acting. The most effective of the different products is reactive rock phosphate from Carolina. It has been recommended for use only in areas with at least 700 mm of annual rainfall and soil of pH 5.6 or less (measured in water). It is also recommended for use on sandy soils so that the loss in drainage water is reduced.

Of the phosphorus applied to the soil as fertilizer, some will likely be lost in the drainage water, some will be removed in plant and animal products, some may be transferred to stock camps, and much will become locked up in the soil in chemical structures which restrict its availability to the plants.

After an optimum level of soil phosphorus has been built up over a number of years the continuing applications should just be enough to maintain that level.

Potash fertilizer is mainly required on lighter soils or where conserved fodder has been removed from the same area frequently. Over-application of this fertilizer should be avoided because, apart from the cost, a higher level of potassium in the pasture plants can contribute to the occurrence of grass tetany in lactating cows.

Nitrogen can mainly boost grass growth, but its use in beef production is usually uneconomic.

Various laboratories conduct tests on soil or plant tissue to evaluate the relative need for different chemical elements. Soil tests can be used to evaluate the need for phosphorus, potassium, and sulphur, and leaf tests can be done to assess the need for sulphur and the trace elements.

In sending samples for analysis it is vital that each sample fully represents the whole paddock. The soil samples are taken with a corer penetrating to 10 cm. The samples should be thoroughly mixed before a subsample is sent to the laboratory.

A useful practice can be to take periodic soil tests on one or two paddocks to monitor the changes in fertilizer needs. With this practice it is vital that the samples be taken from the same parts of the paddock each year.

Recommendations on fertilizer use usually accompany the results of the laboratory tests. The recommendations vary according to the rainfall and soil conditions and how the botanical composition of the pasture is likely to affect the response. They also vary according to the profitability of the enterprise, because the response to increased fertilizer application is at a declining rate.

Fertilizers are usually best applied at the start of the period of pasture growth, and in regions of winter rainfall any improvement in the autumn and winter pasture growth from the application of fertilizers in autumn can be especially valuable. At Rutherglen, superphosphate application increased pasture growth only during autumn and winter.

The main response to phosphate application is greater clover growth, so to obtain a good response to this fertilizer it is necessary that the pasture contains an adequate amount of clover.

There is often not a large immediate response to the application of a phosphate fertilizer, and instead, continuing application usually results in a continuous increase in pasture productivity and carrying capacity. The use of phosphate fertilizer together with sown pasture species has increased the animal-carrying capacity by about three and a half to four and a half times compared to unfertilized pasture.

Pasture Maintenance

The level of fertilizer application and the grazing and fodder conservation practices can greatly affect the botanical composition of the pasture.

The application of phosphate and sulphur stimulates clover growth and the resulting accumulation of soil nitrogen can sometimes stimulate grass growth.

Generally, lighter grazing at any time of the year can enable the more-upright-growing grasses to smother the more-prostrate-growing clover and reduce its presence.

Moderately heavy grazing can favour the competitiveness of the clover, and continuous extremely heavy grazing during winter—such as what mainly occurs with wether sheep—can result in lower plant density and will favour the competitiveness of broadleaved weeds, such as capeweed, which can resist grazing by laying their leaves on the surface of the soil. It can also favour the presence of volunteer grasses, which can set seed quickly and have unpalatable seed heads. Moreover, the high nitrogen levels in the soil accompanying heavy stocking can also favour the growth of broadleaved weeds.

Some soil types can strongly favour some plant species. For example, some of the lighter soils in South Australia strongly favour the competitiveness of Yorkshire fog grass.

Moderate amounts of dead herbage remaining in autumn can help to retain the surface moisture and aid the germination of the seed of annual species, but large amounts of dead herbage can smother clover seedlings especially. Also, the seed of subterranean clover germinates more readily with warmer conditions, so the level of germination and the subsequent clover content of the pasture can vary greatly according to the timing of the opening rains. At Rutherglen, the subterranean clover content of the pasture during winter has generally declined progressively from about 70% to about 10% with a delay in the opening rains between March and June, and with even later opening rains there was little or no clover in the pasture. In the case of subterranean clover varieties with substantial amounts of hard seed, a large amount of this seed can still be viable after at least three years of weather conditions unfavourable for its germination.

An earlier germination also usually results in greater pasture growth until the end of winter.

By preventing the grass from seeding, haymaking on annual pasture can be followed the next winter by a less-dense pasture with little grass. Any effect can be greater with a later time of cutting, and it can also be affected by how much hard clover seed is present and how the timing of the opening rains favours the germination of this seed. On annual pasture at Deniliquin, cutting of annual pasture for hay resulted in reductions in plant density in the following winter of 43% for clover and 71% for grasses. On the other hand, haymaking on clover-dominant pasture did not affect the later pasture production if the cutting was not too late in the season.

At Hamilton, later cutting of perennial pasture for haymaking has tended to result in a greater content of Yorkshire fog grass in the following year.

In Western Australia, spraying the pasture in autumn to remove the grasses reduced the weight gain of wethers by 13% and the daily wool production by 9%. In another experiment, the pasture production was reduced by about 30% with the same practice. The exact effect is likely to depend on the proportion of grasses in the pasture to begin with.

Factors favouring the competitiveness of barley grass are as follows: (*a*) an unpalatable seed head and a capacity to set seed quickly after grazing or mowing, (*b*) non-gathering of its seed by ants, and (*c*) a capacity of the seed to germinate on the soil surface at low moisture levels. On irrigated pasture at Deniliquin, heavy grazing with sheep 20 days after germination almost completely eliminated the barley grass.

In the case of Wimmera ryegrass, the seed heads can be grazed intensely by sheep, and the seeds are harvested by ants. In addition, the seed can germinate relatively poorly if it is not covered.

Some cropping practices can result in more Wimmera ryegrass and silver grass in the later pasture. Silver grass is fibrous and unproductive and can suppress the expansion of other species. If it is present in substantial amounts it is best removed selectively by using the appropriate herbicide.

VOLUNTEER SPECIES

Some volunteer annual species can have obvious undesirable features, such as the damaging seed heads of some volunteer annual grasses and the strongly invasive properties of Paterson's curse. However, there can sometimes be counteracting effects. For example, barley grass can produce a green pick rapidly after the opening rains, and the seed can germinate well at low moisture levels.

Tests have shown that a moderate invasion on an annual pasture by volunteer grasses may have little effect on the productivity of the pasture, especially if the pasture is grazed heavily enough in winter to prevent any frosting of plants such as barley grass. Also, capeweed can grow fairly well, has a favourable mineral content, and is fairly nutritious, but its high moisture content can be a problem in haymaking, and the leaves can rot readily during summer rain.

Soil Acidity

The acidity level of the soil can affect plant growth through its effect on the solubility of some chemical elements in the soil. It can also affect the survival of the rhizobium bacteria responsible for nitrogen fixation in the roots of legumes, and persistent high levels of acidity in the surface soil can lead to acidity levels in the subsoil that restrict root development and plant growth.

Soils are generally acidic in regions of medium to high rainfall, unless there is free calcium carbonate present in the soil, and alkaline in regions of low rainfall.

The acidity level of the soil is measured on the logarithmic pH scale ranging from 1- representing extremely acidic, through 7 – neutral, to 14 – extremely alkaline.

In Australia, the soil pH level is measured most commonly in a solution of calcium chloride ($CaCl_2$), because the alternative method of measurement in water (H_2O) can provide more variable results at different times of the year. The reading in the calcium chloride solution is commonly 0.5 to 0.8 of a unit lower than that in water.

When the measurement is in calcium chloride, a decrease in the pH level below about 6.0 can start to result reduced availability of calcium, magnesium, phosphorus, potassium and molybdenum, and at a pH level of 4.0 or less there may be toxic levels of manganese and especially toxic levels of aluminium Also as the pH level increases above 7.0 there may be reduced availability of iron, manganese, zinc, copper and boron.

Soils become acidic from: a) the leaching of minerals in the drainage water, b) the incorporation of minerals into produce, c) the production of organic acids from the decay of organic matter, and d) any application of raw sulphur or sulphate of ammonia. Also, nitrogen accumulation in the soil from clover growth leads to some of the soil minerals becoming converted to their soluble nitrate form, making them more readily leached.

Plant species differ in their tolerance of different levels of soil acidity. Of the common agricultural plants, lucerne, *Phalaris*, and cereals other than triticale can be relatively sensitive to acidic conditions. Even within the same species there can be differences between varieties in their tolerance of extreme pH levels. Most plants can grow well over a pH($CaCl_2$) range of about 5.5 to about 7.2.

The soil acidity level is reduced by applying ground limestone or dolomite to the soil. Limestone is almost pure calcium carbonate, and dolomite additionally contains different levels of magnesium carbonate. On some soils the addition of magnesium can benefit plant growth. Also, the use of dolomite will increase the magnesium content of the clover on some soils, but on others it may only prevent the slight fall in the magnesium content of the clover that can occur with limestone application. Dolomite can be more expensive than lime.

In an experiment at Rutherglen, when the application of lime or dolomite was compared for improving the magnesium status of beef cows it was found that while the dolomite application prevented the modest reduction in the magnesium content of the clover with lime application it had little effect on the magnesium status of the cows. This was because the magnesium status of the cows was mainly affected by the influence of the weather conditions at the opening rains on the later clover content of the pasture.

Samples of limestone and dolomite are assessed on their neutralizing value, particle size, and water content. Finer grinding results in faster action, and the lime or dolomite needs to be worked into the soil to get a reasonably fast response.

Because of the relatively good response of cereals, the financial return from liming is usually quicker with cropping than with continuous pasture.

A recommendation by the New South Wales Department of Primary Industries is that it can be desirable to maintain a pH($CaCl_2$) level of about 5.5 in the top 10cm of soil and about 5.2 in the next 10cm layer.

The slow movement of lime in the soil makes it more difficult to change the pH level in the lower layer.

Pasture Restoration

Before starting expensive cultivation and reseeding to restore a pasture, it is important to consider why the pasture deteriorated in the first place and whether any actual improvement will be great enough and prolonged enough to justify the cost.

If a pasture has a uniform representation of desirable species, a quick and adequate improvement can often be obtained by applying fertilizer and using different grazing intensities at different times of the year to favour the competitiveness of one or more of the plant species. Herbicides can sometimes also be used to an advantage, and their use can be necessary for controlling some broadleaved weeds such as Paterson's curse.

The germination of seeds is usually better if the seeds are buried, but the content of subterranean clover in a pasture can sometimes be improved by spreading seeds on the soil surface just before the opening rains. Having a light cover of dead herbage to conserve moisture can also be beneficial. Surface-spreading of grass seed in an annual pasture is usually of little benefit.

Drainage

Water-logging can greatly reduce the pasture productivity and greatly affect the relative persistence of the different plant species. Highly effective subsurface drains are usually grossly uneconomic for use in beef production, but simple surface drains that remove surface water can often be highly beneficial.

Lucerne

Lucerne is sometimes grazed by beef cattle. New South Wales results indicate that this species should be grazed rotationally, with the grazing occurring when the plants are about 15 to 30 cm high and no later than the start of flowering. Precautions need to be taken against bloat when cattle graze lucerne.

Lucerne can also be used for haymaking, but if it is not cut when immature, the resulting hay can be highly fibrous and of modest feeding value.

Conclusions

To achieve productive pastures, it is important to achieve good establishment of appropriate species to begin with, fertilize adequately, and use grazing and fodder conservation practices that will favour the persistence of a favourable botanical composition. In some situations, the use of the appropriate fertilizer and grazing practices can be a highly effective method of pasture regeneration.

CHAPTER 4

Grazing Management

The way pastures are grazed can greatly affect the labour requirement and profitability of beef production.

The main factors in grazing management are the stocking rate and level of supplementary feeding. Hay is the most common supplementary feed, and silage and grain are of lesser importance. Supplementary feeding may be used frequently to try to improve the pasture utilization and animal performance or used only in such situations as years of poor pasture growth.

In this chapter, the effects of stocking rate and supplementary feeding will be examined for steers and beef cows, and then other aspects of grazing management will be addressed.

STEERS

The response of steers to stocking rate and supplementary feeding was examined over four years at Rutherglen, using steers bought at weaning in early summer and kept for a year before being slaughtered and replaced.

At each of the five stocking rates, hay was provided as required during late autumn and winter at the levels of N (no hay), M (enough to prevent any weight loss), and G (enough to gain weight at 0.45 kg a day). The pattern of response for gross margin per hectare (the gross return less variable costs) did not vary when assessed according to the differences in the costs and prices between years, and the values reported for 1988 prices and costs were as follows:

Level of Hay-Feeding	Margin/Hectare ($) Steers/Hectare				
	0.7	1.1	1.4	2.1	2.8
N	78	92	97	59	−36
M	−	−	−	−	−49
G	−	−	72	35	−47

As shown above, the most profitable practice was to use the middle stocking rate and to not provide hay. A similar pattern of response applied on perennial pasture in Western Victoria. In each instance, the most profitable treatment was to stock at the heaviest rate that resulted in the steers being finished to market requirements by the end of the annual period of pasture growth without hay-feeding. Also, in two years at Hamilton the total weight gain of weaner steers did not benefit from making hay on a tenth of the plot area during spring and feeding it back during autumn or during autumn and winter. Furthermore, a computer model developed by a CSIRO scientist indicated that frequent use of hay is not an effective way of increasing beef production.

The Rutherglen results were obtained using hay of good quality by district standards. Also, the hay was not made on the plots, so the results do not reflect the harmful effect that haymaking can often have on the later plant density and productivity of annual pasture.

Later results obtained at Rutherglen with similar steers and hay to that described above indicate some of the factors that affect the respond to hay-feeding. In this instance, hay-feeding during late autumn and winter was compared with no hay-feeding at two stocking rates, and the steers were kept until the end of spring. The liveweight changes were as follows:

Steers/Hectare	Hay/Steer(kg)	Weight Change (kg)		
		Autumn/Winter	Spring	Total
1.8	0	17	105	122
1.8	685	59	77	137
3.7	0	-49	98	48
3.7	699	58	63	121

As shown above, the response during late autumn and winter to a similar, substantial amount of hay was better on the poorer pasture conditions at the heavier stocking rate. Also compensatory growth during spring in the nil-hay steers reduced the earlier benefit from hay-feeding, so that by the end of spring a substantial benefit remained only at the heavier stocking rate.

This variable response to hay-feeding clearly indicates that at least at the lighter stocking, hay-feeding must have resulted in reduced pasture consumption. Because of the amount of pasture involved, this effect was not evident from just looking at the pasture and could only be detected by careful measurement. Results from other locations have also shown reduced pasture consumption with supplementary feeding, even on poor pasture conditions.

The quality of the hay will affect the degree of response that can be obtained from supplementary feeding. Tests on farmers' hay in Western Victoria and at Rutherglen as well as observation of when pasture is being cut for haymaking, indicated that much of the hay made on farms is only of mediocre quality, resulting in hay that is mainly suited to preventing a weight loss under very poor pasture conditions rather than for promoting weight gain. The main factor contributing to poor hay quality can be a delayed cutting time because of unsettled weather conditions.

Both at Hamilton and Rutherglen, terminating hay-feeding at different times during winter to try to induce the steers to make better use of the winter pasture before the start of faster pasture growth in spring was of no apparent benefit.

With sheep, a worthwhile response to hay-feeding has also generally been obtained only at stocking rates heavier than the most profitable one, and having to be at such a heavy stocking rate adds to costs.

There is no benefit in giving hay to cattle during winter just as a source of fibre. Cattle have soft dung when grazing winter pasture, but if their dung is watery, worm burdens should be suspected. Also, in Western Australia, hay provided without a magnesium supplement was inadequate for preventing grass tetany, and in Victoria, providing hay on its own was ineffective in preventing bloat in dairy cows.

At Rutherglen, paddock-stored hay was tested as a means of improving the growth of weaners over the period of dry pasture. It became evident that because of the weather conditions at the time of haymaking, hay of good enough quality could not be made consistently enough for the practice to be profitable, and that much or all the benefits to animal performance during summer and early autumn could be lost during the following late autumn and winter because of the reduced production from annual pasture that often follows haymaking.

This last practice of using paddock-stored hay was also used at Rutherglen with dry-land lucerne. The paddock was closed to grazing at the start of spring and the lucerne cut twice. It was evident that the best practice was likely to be to delay closing the paddock so that immature lucerne would be available for cutting when the drying conditions had become reliable. The daily live weight gains achieved in weaner steers on the first and second cuts were 0.1 and 0.7 kg respectively.

Overall, it is clear that the most profitable method of grazing management for steers to be finished for sale at the end of spring is generally to stock at a level that enables them to be finished for slaughter at that time with only an occasional need for supplementary feeding.

Beef Cows

Results obtained at Hamilton and reported in the chapter on milk production, indicate that the most profitable stocking rate for beef cows is generally the heaviest at which a high level of fertility can be maintained without frequent supplementary feeding. A similar effect has been evident in results obtained at Canberra and in America.

When beef-breed cows of good milking capacity were stocked at this level at Hamilton, the calves were of vealer standard at weaning. A similar result was obtained at Rutherglen.

There is only a limited amount of information available on the effect of supplementary feeding of beef cows under the pasture conditions applying in southern Australia. However, supplementary feeding is likely to be less effective with beef cows than with steers, because both cows and calves are being fed just to improve calf growth, and the milk production of beef cows can respond only moderately to improved nutrition, as is described in the chapter on milk production.

The main benefit of supplementary feeding for beef cows is likely to be obtained in years when the pasture growth is so poor that cow fertility as well as calf growth is likely to benefit.

Supplementary feeding practices were evaluated on annual pasture at Rutherglen with Angus cows calving in early summer and the calves weaned at an average age of 10.5 months at the end of spring. These animals were expected to be especially responsive to supplementary feeding because of the long period of dry pasture they would experience after calving. Also, in deciding the treatments it was known that calf growth can respond little to the pasture conditions until after about 2 to 3 months of age.

The intended treatments were to supplementary feed the cows or the calves on dry pasture until the start of pasture growth, but because of a late germination and poor winter pasture conditions the treatments were continued until the end of winter. Moreover, a treatment of providing hay to the cows and calves was added.

The treatments eventually applied were as follows: (*a*) no supplementary feeding, (*b*) a supplement of 1 kg of rolled wheat and 0.5 kg of rolled lupins a cow daily from late February to boost milk production, (*c*) a creep feed of 2/3 rolled oats and 1/3 rolled lupins to the calves from late February, and (*d*) as much hay as could be provided three times a week with little wastage from late May.

The amounts of supplementary feeding provided and the average calf carcass results were as follows:

Treatment	Supplement/ Cow and Calf (kg)	Increase in Calf Carcass Weight (kg)	Return/Tonne of Feed ($)
Cow supplement	315	14	84
Creep feed	300	19	120
Hay	1,330	22	31

None of the supplementary feeding responses was great enough to be significant, but the above economic assessment indicates the situation that would have applied if these results are regarded as significant. Accordingly, the best economic return from supplementary feeding was from the creep-feeding, but even it would not have returned the cost of the supplement. Also, in a farm study in central Victoria with a breeding herd in which the progeny were retained for fattening, the response to hay-feeding was uneconomic.

Results from Hamilton and elsewhere have indicated that calves born in autumn or late winter are likely to respond markedly to creep-feeding only after the pasture dries off in late spring or early summer. In this situation, it also seems that about 7.5 to 9.5 kg of oats are likely to be required for each kilogram of improvement in weaning weight. Some of the calves can experience acidosis.

From several sets of overseas results, it is evident that the economics of creep-feeding is likely to be better if the calves are sold at weaning rather than being retained, because the benefit from creep-feeding can decline progressively after weaning.

Overall, it seems that the most profitable grazing method for beef cows is usually to stock at a level that results in consistently high fertility levels with only an occasional need for supplementary feeding. Also, it seems that creep-feeding of calves can sometimes be effective, but the economics of the practice needs to be assessed carefully.

Stocking Rate

There is no need to try and be precise in deciding the stocking rate. Pasture conditions vary from year to year unpredictably, and it is usually impractical to change animal numbers accordingly. Also, near the most profitable point, the profit level changes little with a modest change in stocking rate. Therefore, all that is required is to use a stocking rate that is likely to be about the most profitable one on average over a number of years.

Recording and refining what appears to be the best stocking rate for each paddock can be a useful management practice, especially where there is a rotation over the years between using the paddocks for cropping and grazing.

Pasture Utilization

While the stocking level should be decided primarily on meeting the nutritional needs of the animals, it is still important to be alert to possible opportunities for improving the level of pasture utilization profitably. In the case of beef cows, the calving season used can affect the weight of calf weaned and the consequent amount of calf weaned per hectare. Also, in the case of steers, a method used in an experiment in Canberra and later on a farm at Rutherglen was to purchase weaner steers in late winter and keep them for about 15 to 16 months, so steer numbers were doubled during the period of rapid spring growth. By better balancing the animal numbers to the seasonal pasture conditions this practice can result in better weight gains combined with good pasture utilization.

Another benefit of this practice at Rutherglen compared to the common practice of buying weaner steers in summer or early autumn, was that if it became evident by late June that a year of poor pasture conditions was likely to follow, the steers were usually heavy enough to be sold into a feedlot at higher prices than could be obtained by selling into a store market. The long-term weather records show that for Rutherglen and much of north-eastern Victoria, a year of very poor pasture conditions occurred on average about every four years, and that only in one year was a germination after the end of June followed by much pasture growth.

ROTATIONAL GRAZING

The merit of rotational grazing has been evaluated extensively in different countries.

Intense interest in the subject followed a finding that pasture plants generally grow faster up to a certain height because of a greater leaf area to intercept sunlight. However, other factors affect plant growth, such as the time tall plants take to recover after being grazed short.

Now it is known that rotational grazing is likely to be of substantial benefit only if the pasture plants would be seriously harmed by continuous grazing.

Lucerne usually needs a spell of at least six weeks after grazing for good plant survival.

It t has also been reported that the persistence of winter-growing varieties of *Phalaris* can be reduced by continuous heavy grazing by sheep during winter, and with all varieties of *Phalaris*, continuous heavy grazing during spring can reduce the development of basal buds that aid plant survival. However, a problem with lax grazing of *Phalaris* during spring can be the development of large clumps of unpalatable herbage.

The plant density and productivity of annual pasture can sometimes be reduced to some degree by continuous extremely heavy stocking with sheep during late autumn and winter, such as is likely to apply mainly with wethers. However, in two experiments with wethers in Victoria, rotational grazing or feeding the animals hay off the pasture in autumn until the pasture had become established was not profitable. Also, in one result, the growth of prime lambs was greater with set-stocking than with rotational grazing, and in an American experiment, the growth of steers was 5 to 8% better with set-stocking than with rotational grazing.

Rotational grazing has also not benefited worm control, and at Canberra rotational grazing of *Phalaris* resulted in more sheep deaths from *Phalaris* toxicity because of hungry animals being put on to toxic pasture.

PADDOCK SIZE

At one stage there were suggestions that smaller paddocks might enable better utilization of the pasture and thus better animal performance. However, it was found that reducing the paddock size below what is required for efficient management of the farm is likely to increase costs and not benefit animal production.

PREFERRED ENTERPRISE

Breeding can be more attractive than rearing and finishing, partly because it is self-replacing. However, the results of a few studies have shown that rearing and finishing can be the simpler and more profitable enterprise if well managed. Some of the disadvantages of breeding are failure to calf, a comparatively high death rate for cows and calves, bull costs, higher veterinary costs, greater labour requirement, and depreciation of the cows and bulls with age.

CROPS

In Western Victoria, the daily weight gain of steers stocked on turnips from 12 February until 21 May was 0.65 kg on one variety and 0.83 on another.

At one stage it was thought that sowing oats into dry paddocks in autumn could result in better herbage production during late autumn and early winter. However, it was eventually found that while oats usually resulted in taller early herbage the plant density was relatively low, resulting in no increase or some reduction in the total weight of herbage produced compared to that from established pasture.

Out-of-Season Fattening

In each region, beef prices are usually highest for the year just before the start of the period of rapid pasture growth. This can provide an incentive to fatten cattle for sale at that time of year. However, in doing so, it is important to consider any additional cost or lower production from such things as (*a*) a greater need for supplementary feeding, (*b*) likely higher costs for holding animals for long periods with little or no weight gain, and (*c*) possibly poorer pasture utilization and lower overall animal production from a poorer balance between animal numbers and the seasonal pasture growth.

Older, heavier animals are usually easier to fatten out of season.

Grain-Feeding

Grain can be an alternative supplementary feed to hay. It is higher in energy than hay but is usually more expensive. Cattle must also be introduced progressively to digesting grain if acidosis is to be avoided.

The few results for grazing steers given grain have been so variable that it is difficult to draw conclusions other than that the response is likely to be better with poorer pasture conditions and with a crude protein level in the diet of no less than about 10 to 11%.

To fatten cattle, a more reliable practice than providing grain at pasture can be to shut the animals in a yard and feed them to appetite on hay and grain. There are various booklets describing methods for this kind of feeding. However, intensive feeding on a farm is likely to be profitable only when the beef-to-grain prices are relatively favourable and when only a short period of feeding will be required. For occasional use, simple methods of intensive feeding can be adequate.

A general recipe for intensive feeding is as follows:

a) Use a hay–grain ration that will contain at least 12% crude protein for younger cattle and 10% for older cattle.
b) Condition the animals to grain digestion over three weeks, then provide as much grain twice a day as the cattle will eat without digestive upset, and make hay freely available at the other times.
c) All grain except oats should be coarsely crushed, and grains other than oats will generally provide a substantially faster weight gain.
d) Crushed limestone and salt should be added to the grain at 1% each.
e) Urea added at 1% to the grain, with sulphur added at 5% of the urea, will boost the crude protein content of the grain by about 2.4%.
f) The urea should be dissolved in water and the solution sprayed on to the grain, thoroughly mixed in, and allowed to soak in.

When steers in Hamilton were fed 5.4 kg of oats a head daily plus pasture hay to appetite for 96 days, the average values were 1.1 kg for the daily liveweight gain and 10.9 mm for the thickness of fat cover on the carcasses. The crude protein levels were 11.2% for the oats and 11.6% for the hay, and the quantities of feed provided were 703 kg of hay and 436 kg of oats a head. The conversion ratio was 10.7 kg of feed for each kilogram of live weight gain.

Queensland results indicate that steers may respond to crude protein levels in the diet up to about 12 to 14%, but the degree of response may not be great enough to make such levels of protein economic. Moderate differences in crude protein level often have only a slight effect on animal performance.

Lucerne hay can be used to increase the protein content of the ration, but even when it forms only 10% of the ration it can sometimes cause bloat.

Various laboratories will evaluate feeds. Among the values reported will be those for metabolizable energy (ME) and crude protein. A higher ME value indicates a higher energy value for animal feeding.

With all cattle to be fed high levels of grain, it is highly desirable that they be vaccinated against enterotoxaemia. Also, routine vaccination of grazing cattle with a five-in-one or seven-in-one vaccine against clostridial diseases can prevent the occasional death from causes that can be difficult to identify.

Conclusions

For steers to be finished for market by the end of spring and for beef cows, the most profitable grazing method is usually to set-stock at levels that result in an adequate finish in the steers and a high fertility level in the cows, with only infrequent need for supplementary feeding. Within this overall practice there can be some opportunity to further refine the management practices.

The merit of supplementary feeding is reduced by the cost, the modest nutritional value of some supplementary feed, the reduction in pasture consumption with supplementary feeding, and the opportunity for later compensatory growth in animals that were not on supplementary feeding. A further effect applying mainly on annual pasture can be damage to the later botanical composition and productivity of the pasture from haymaking.

Grain can be fed to cattle at pasture or in yards, but the economics need to be carefully assessed, and practices need to be used to protect against acidosis.

CHAPTER 5

Grazing Behaviour

Knowing how cattle graze can help in understanding what affects their performance.

TEETH STRUCTURE

Calves are born with 20 teeth, or they emerge shortly afterwards. Eight of these teeth are sharp incisor teeth located at the front of the bottom jaw, and they bite on to a hard dental plate on the top jaw. At the back of each jaw on each side are broad molar teeth used in grinding. Later, these initial teeth, or milk teeth, are replaced by larger permanent teeth, and molar teeth are added, bringing the total number of teeth to 32.

The time of replacement of the temporary incisor teeth at the front of the bottom jaw is used to estimate the age of the cattle. These teeth are replaced progressively in pairs, starting with the middle pair and moving outwards. The age at replacement is about 18–24, 24–30, 36, and 42–48 months for the successive pairs of teeth, but there can be substantial differences in the replacement time between animals.

Later in life, the teeth can become worn and loose, just as is the case with sheep, and when this stage is reached the animal's grazing ability is reduced. At what age this occurs is likely to vary according to the grazing conditions the animal is experiencing. The effect of deteriorating teeth on animal performance is likely to be greater with shorter pasture.

Pasture-Grazing

When cattle graze on pasture they keep moving forward with their heads moving from side to side. If the pasture is long they use their tongues in a sweeping motion to direct pasture towards their mouths, but if the pasture is short they use their tongues only to clear the harvested pasture from the front of their mouths. The tongues of cattle are highly muscular, and when extended they become narrower and more flexible and can grasp strongly.

In consuming pasture, cattle clasp the pasture between their incisor teeth and dental pad and break it off with a short, sharp tug of their heads. The pasture is swallowed almost straight away or given a quick chew between the molar teeth. Saliva flowing into the mouth from salivary glands is mixed with the pasture.

During periods of the day, boluses of the feed consumed are regurgitated for further chewing, a process called rumination.

According to results obtained with sheep, cattle are likely to use their senses of smell, sight, and taste in selecting herbage. Some plants, such as docks, can be acceptable at some stages of their growth but not at others. Cattle generally select a diet that is more nutritious than the pasture as a whole. This could be due to the cattle avoiding patches of the pasture that are more fibrous and tougher to harvest, or their practice of grazing downwards from the top of the pasture may just result in the consumption of a higher proportion of leaf to stem than is present in the pasture as whole. There has been no indication that cattle can refine their selection practices according to their relative need for different nutrients, such as minerals.

Cattle on average spend about a third of the day grazing, a third ruminating, and a third resting. However, there is great variation in these values. The reported grazing times have been generally between about 5 and 13 hours a days. The actual grazing time seems to depend on the appetite of the animal and how readily it can be satisfied. Fatigue seems to impose an upper limit on the grazing time.

How much pasture cattle consume during grazing depends not only on the time spent grazing but on the grazing vigour and amount of pasture harvested in each bite. It has been observed that hungrier cattle bite faster, and how much pasture is consumed in each bite depends on the height and density of the pasture. The bite size can be relatively small on both short pasture and on tall pasture with an open canopy. Supplementary feeding has been observed to reduce both the grazing vigour and time spent grazing.

During the cooler months of the year the main grazing periods during the day are just after sunrise, about the middle of the day, and just before sunset. In days with short periods of daylight the grazing can be almost continuous during the daylight hours.

At times of the year with very hot weather the main grazing periods are just after sunrise and during the last few hours before sunset.

Variable results have been reported for the amount of night-time grazing in temperate climates. It seems that there can be some grazing during the night, especially on moonlit nights and during summer, but there can be little grazing on cold nights, especially on frosty nights. In an American study, the times spent grazing during the day and during the night were 6.45 and 4.53 hours in August and 7.3 and 3.30 hours during the colder month of October. Several sets of results have indicated that there is generally little grazing through the night when the maximum day temperature is below about 15 °C, and when the maximum day temperature is 25 °C or over the amount of grazing time at night can be up to more than half that for the day as a whole.

There can be reduced daily grazing with the persistently hot, humid weather found in the tropics, but in more-temperate climates the effect of cool nights can largely counteract the effect of hot days on the total daily pasture consumption.

The grazing time can also vary with animal age. In one instance the average grazing times were 8.5 hours at 8 to 17 months of age, 7.4 hours at 17 to 24 months of age, and 6.9 hours at 3 years of age.

In one set of British results there was a 21% degree of variation in the grazing time between individual cows in a dairy herd. The reasons for this effect were not identified, but they could have included differences in appetite because of differences in milk production. Other British results have indicated that the energy expenditure of grazing may increase the nutrient requirement by about 20 to 40% compared to that of cattle hand-fed in a yard.

Conclusions

Apart from the weather, whose effects are described in another chapter, the main influences on cattle-grazing practices are the height and density of the pasture and the appetite of the animals. The seasonal changes in climate and daylight hours can affect the times of day cattle graze.

CHAPTER 6

Climatic Effects

The climate can affect animal performance directly and through its effect on pasture conditions.

BODY TEMPERATURE

The normal body temperature of cattle is 38.3 °C. It can vary moderately from this level in very hot or very cold weather, but if it changes too much the bodily processes of the animal can start to malfunction, and the animal can die. In one report from the tropics the body temperature of some cattle commonly rose as high 40.5 °C during the day.

Heat is generated within the body from the chemical reactions involved with feed digestion, body metabolism, and movement. Heat is also obtained from the sun's rays and from the air during extremely hot weather. The more feed consumed and the more active the animal, the greater is the amount of body heat generated.

To help maintain body temperature, feed consumption can fall during hot weather and rise during cold weather. The degree to which feed consumption will be affected by a change in temperature will depend on how much feed is already being consumed initially and how its nutritional value is affecting the amount of body heat being generated. Only modest amounts of heat are likely to be generated when consuming dry pasture, whereas large amounts of heat are generated in feedlot cattle consuming large amounts of grain. With feedlot cattle it has been recorded that the feed consumption may fall when the maximum daily air temperature exceeds about 25 °C, and increase when it falls below about 15 °C. In one study it was concluded that the feed consumption of feedlot cattle is likely to decline by about 10% when the average maximum temperature during the day reaches 25 to 30 °C and increase by about 3% when the average maximum temperature for the day declines to 15 to 5 °C. Deep mud and cold wind and rain further reduce feed consumption, and the provision of properly constructed shade can result in improved feed consumption. Also, cool nights help to compensate for the effects of hot days on the appetite of the animal.

During winter in a temperate environment the feed requirement for cattle to maintain body weight will rise to some degree with the cooler weather, but the effect on animal weight change is likely to vary according to how the pasture conditions enable the animal to compensate by increasing consumption. On annual pasture at Rutherglen the change in live weight of steers relative to pasture height did not differ significantly between winter and spring.

If the feed consumption is inadequate to maintain body temperature the animals will have to draw on their body reserves of energy for this purpose, and will lose weight.

Heat can be lost by sweating, evaporation from the lungs, the evaporation of any moisture falling on the skin, together with radiant and convection heat given off by the skin. Radiant heat is the kind given off by a heating element such as that of an electric radiator, and convection heat is the kind given off when air moves over a hot object.

Drier air and faster air movement increase the evaporation speed, and colder air and faster air movement increase the loss of convection heat. Also, rain has a direct cooling effect, and when water evaporates from the skin heat is withdrawn from the skin to convert the liquid water into water vapour. Thus, overall, a strong, cold wind combined with rain can have an especially severe cooling effect.

During cold, wet, windy weather, cattle seek protection from the wind, and if this is not available they will stand in a huddle with their tails into the wind to reduce the direct force of the wind on their bodies. They will also reduce their grazing time during such conditions, but if these conditions last for only a few days the cattle will soon compensate. Also, under stressfully cold conditions cattle can shiver to help to maintain body temperature, but this can only continue for a short period.

Larger animals have a smaller surface area relative to their weight, which results in less surface heat loss relative to their weight. Also, subcutaneous fat acts as an insulating layer. Thus, larger and fatter animals have greater cold tolerance but lower heat tolerance. In Canada, the large European beef breeds have proved to be unsuitable for use in very cold regions because of their lack of fat cover. On the other hand, heavy, fat animals generating substantial amounts of heat from the consumption of high-energy rations in a feedlot can become severely stressed during very hot weather. In one instance, at an air temperature of 35 °C, the number of breaths per minute was 60 for fat cows compared to 49 for leaner cows. In another feedlot experiment conducted over three and a half months, when the daily temperature ranged from 14.5 to 20.1 °C minimum to 26.9 to 33.3 °C maximum, the overall average response per animal from providing shade were increases from 8.80 to 9.46 kg in the daily consumption of feed dry matter and from 1.41 to 1.60 kg in the daily live weight gain.

One claim has been that for an artificial shade structure to be effective it needs to be of reflective material held at least 3 m from the ground, and provide at least 6 sq. m of cover for each adult animal. In American feedlots there has been more benefit from providing shade in hot, dry climates than in hot, humid climates, because of the greater relative importance of radiation on body temperature in a drier environment.

In one assessment, the energy requirement to maintain weight in cool weather was 7 to 12% greater for beef x dairy breed cattle than for beef-breed cattle because of the greater surface area relative to live weight in the crossbreds, as well as a larger digestive tract and thinner fat cover.

Breed Effects

The two main species of cattle are the *Bos taurus*, which are adapted to temperate climates, and the *Bos indicus*, which are adapted to tropical climates.

Attributes of the temperate breeds include thick skins and coats. These coats vary in thickness according to the seasonal temperatures, and the pile can rise during very cold weather to increase the insulating properties. Also, except for a few cases such as the Jersey dairy breed, the temperate breeds have few sweat glands. Therefore, they mainly lose heat by evaporation from the lungs and by radiation and convection from the skin. During very hot weather they can start to pant to increase the evaporation from their lungs.

Evaporation from the lungs is a much less effective cooling method than sweating, and this limits the heat tolerance of temperate breeds. However, temperate-breed cattle can become adapted to some degree to high temperatures.

Another control mechanism is that during hot weather more blood can be directed towards the lungs in temperate breeds and towards the skin in tropical breeds, and with both breed groups blood can be withdrawn from the skin during cold weather.

In breeds with extreme cold tolerance, such as the West Highlander, the coat consists of two layers—an outer layer of long fibres that shed water and an inner layer with better insulating properties.

Even within the same species of cattle there can be small differences which make one breed better suited than another to particular conditions. In northern Australia, when British breeds were used more extensively than at present it was claimed that Hereford cattle were comparatively well suited to timbered countryside and Shorthorn cattle to the more open plains.

During very hot weather cattle also seek shade to protect themselves from the sun's rays, and they will change their grazing times to avoid grazing during the hottest part of the day. In temperate climates, cool nights usually cancel out the effect of hot days on daily pasture consumption. Also, during summer there will only be a modest amount of heat generated from consuming the dry pasture.

Attributes in tropical breeds that provide them with better heat tolerance are a) a substantial capacity to sweat, b) a thin skin which aids heat transfer, c) a shiny skin that better reflects the sun's rays, d) a large dewlap and an angular body which increase the surface area, and e) about a 12–15% lower level of metabolic activity, which reduces the amount of body heat being generated.

Tropical breeds are also comparatively good at digesting herbage with the higher fibre content and lower protein content common in the tropics. Moreover, they are highly resistant to eye infection and substantially resistant to the skin parasites common in hot, humid climates.

While these attributes of tropical breeds can be beneficial in the tropics, they can be a disadvantage in more-favourable situations. In the subtropical environment at Grafton in northern New South Wales, the best cattle to use proved to be a tropical breed on poor, unimproved pasture; a temperate x tropical breed on semi-improved pasture; and a temperate breed on highly improved, irrigated pasture.

Conclusions

Overall, in southern Australia the main effect of the climate on the performance of grazing cattle is likely to be through the effect on the quantity and quality of the pasture on offer, and any more-direct effect is likely to occur mainly with lean animals during winter and with those in coastal regions with long periods of persistent rain and cold wind during winter.

CHAPTER 7

Fodder Conservation

Hay is the main conserved fodder used in beef production, but silage is sometimes also used. The quality of the feed affects the level of animal performance that can be achieved.

Haymaking

The quality of the hay that can be made depends greatly on the maturity of the plants at cutting. Lower maturity can be achieved by cutting the pasture earlier in spring or by shutting up paddocks for shorter periods. Cutting earlier can be tricky because of unreliable weather conditions. Making hay from less bulky pasture will increase the cost per tonne, but this may be unavoidable if better quality is required.

Weather forecasts are now available for a week ahead, and they can be of great benefit in haymaking.

Continuing respiration in the pasture plants after cutting results in nutrient loss. Faster drying reduces this loss. The drying is faster if the sward is lighter and the distribution of the herbage within the sward is more even. Tedding as soon as possible after cutting to fluff up the herbage can substantially improve the drying rate, especially if the mower leaves a swath of uneven thickness. Crushing or crimping at cutting can reduce the drying period by up to about two days. This practice has proved to be mainly of benefit with succulent, immature herbage and with material with coarse stems, such as lucerne. Capeweed can be slow to dry because of its thick stems and high moisture content.

The drying rate is affected by the relative humidity and temperature of the air and the wind speed. The drying rate is rapid to begin with and then slows down.

Rain falling on the cut herbage can result in some loss of soluble nutrients. With repeated wetting the degree of effect can depend more on the number of wettings rather than on the total rainfall. Wet conditions that last long enough to cause rotting can be especially damaging. The losses of dry matter with wetting can increase as the pasture becomes drier. In one instance, the loss of dry matter from applying 20 mm of artificial rain was 1% when the cut pasture contained 80% moisture and 8% when it contained 26% of moisture.

The leaves are the most nutritious part of the plant, and any leaf loss during haymaking can be detrimental. The loss of leaf in haymaking can be affected by the dryness at baling, the roughness of raking, and how well the sward feeds into the baler. The leaf loss can be especially great with lucerne and clover.

Any heating or moulding of the hay after baling can greatly affect the nutritive value, and baling the hay when it is too wet can sometimes lead to spontaneous combustion later. Also, if moist hay stays at a temperature of about 40 °C for a prolonged period, spores can form which can cause lung problems in humans and animals.

An increase in the clover content can result in improved mineral and protein contents, and in the case of perennial pasture it can also result in some improvement in the overall digestibility of the hay. However, on annual pasture the clover content may sometimes have little effect on the digestibility.

At Hamilton, in a year in which hay could be made from immature pasture under perfect drying conditions, the daily weight gain achieved with steers in yards was 1.25 kg a day, but hay of anything like that quality can usually be made only rarely. Most hay made is only of modest-to-low nutritional value. Nevertheless, an observation at Hamilton was that, despite the greater risk of spoilage with early-cut hay, its nutritional value was never below that of late-cut hay. Also, at Hamilton the feeding value of cereal hay has been about the same as that for late-cut pasture hay.

The use of large round rolls has reduced the cost of haymaking, but it has not reduced the difficulty in obtaining a worthwhile animal response to the use of hay. Also, if paddock-stored hay is not used within a few months of making there can be substantial rotting. In South Australia, the loss of dry matter from fodder rolls over three, six, and nine months was 17, 26, and 45% respectively when the rainfall in these periods was 165, 319, and 428 mm.

Silage-Making

Silage is an alternative supplementary feed to hay. It has the advantages over hay that it can be made earlier in spring when the pasture is more nutritious, and it can be made under more variable weather conditions. It can also be a suitable feed reserve for use in a drought. However, it can be less convenient to use than hay, and when provided to appetite the consumption of dry matter can be lower, probably because of the acidity level and the breakdown of protein during the fermentation. Moreover, the loss of dry matter with some methods of storage can sometimes be high, and losses from rotting of up to about 44% have been recorded with wedge-shaped stacks of 50 to 60 t.

In Britain, compared to fresh pasture, the average decline in the dry matter intake was 18.5% with hay and 35.8% with silage, and the range of decline in the digestibility compared to the original pasture was 0 to 15% for hay and 0 to 10% for silage.

As with haymaking, the quality of the product depends greatly on the nutritive value of the pasture at cutting. Making silage earlier in spring or after shorter periods of paddock closure can result in a better-quality product. The two most important factors in silage-making are the maturity of the herbage and the standard of fermentation. In one instance, when the paddock was shut up for 12 or 6 weeks for silage-making the resulting average daily liveweight gain achieved in cattle were 0.19 and 0.56 kg with the respective silages.

Preservation in silage-making is achieved by the fermentation of sugars to organic acids. The pH of fresh herbage is generally about 6 and it has to decline to about 4 before full preservation is achieved. Grasses are much easier to ensile satisfactorily than legumes because of their greater sugar content, and the chemical composition of legumes, retards a change in the pH level. Also, the sugar content of pasture plants can rise during the day on sunny days.

A higher moisture level in the herbage requires a greater amount of fermentation to achieve the required pH level, resulting in a loss of energy. It can also result in an undesirable butyric acid kind of fermentation compared to the desirable lactic acid kind of fermentation, as well as resulting in greater breakdown of protein and greater loss of nutrients as effluent. On the other hand, if the herbage is too dry there can be overheating and a reduced nutritive value. Coarse herbage, such as cereal crops, can also be more subject to overheating because of greater air entrapment.

Wilting the herbage to a dry-matter content of about 25 to 30% can result in a more reliable fermentation and a better nutritional value. A rough guide to this moisture level is to chop samples of the pasture into 1–2 cm lengths and see how much moisture is expelled when a ball of this material is squeezed tightly by hand and released quickly. At the desired moisture level the ball will hold its shape and there will be no free moisture expelled, but the hand will be moist. If moisture is expelled or the ball starts to fall apart slowly the material is wetter or drier than desirable. Another test that has been proposed is to see if any moisture can be expelled when a sample of pasture is twisted tightly.

Herbage is now sometimes also wilted to a moisture content of about 50%, formed into rolls, and wrapped in plastic to achieve complete exclusion of air. This method is almost certain to be uneconomic for use in beef production in Australia.

Chopping the herbage before ensiling can result in better compaction and better control over the fermentation temperature, resulting in a better product. It also leads to easier removal of the silage. Machines are available which can chop the herbage to different lengths. British recommendations are that the maximum chop length can be increased progressively from 2.5 to 20–25 cm with an increase in the dry-matter content of the herbage over the range of 20 to 30%. The shortest chop length is also recommended for use with tower silos and where the animals will self feed.

In Western Australia, when subterranean clover/perennial grass pasture was single-chopped to 20–25 cm, double-chopped to 8–12 cm, or precision-chopped to 3–5 cm for silage-making and the silage was self-fed to steers, the average daily liveweight gains were 0.49, 0.57, and 0.57 kg on the respective silages.

In several reports, chopping the herbage has resulted in a substantial increase in silage consumption but little increase in animal performance. The reason for this effect has not been clear.

In the past, molasses was commonly added to the pasture when it was being ensiled to increase the sugar content. However, this practice proved to be unreliable because of uneven distribution of the molasses. Formic acid is commonly used overseas when making silage from unwilted herbage. The benefits can be a more rapid decline in the pH level and less protein breakdown.

It also used to be common to roll the silage for several days with a tractor to try to increase the compaction and moderate the fermentation temperature. Now, however, it is known that this rolling has a bellows effect, drawing in more air and continuing the fermentation. The current recommendation is that the rolling during filling can be adequate, and that the silage should be covered straight away with a plastic sheet to stop hot air rising through the silage and sucking fresh air in. Covering the silage with plastic and fixing the plastic tightly in place also stops rain infiltration and reduces surface rotting.

The relative feeding value of hay and silage depends on how well each product can be made, and this will largely depend on the weather conditions. At Hamilton, when hay and silage were made at the same time and both were fed in unrestricted amounts to steers, the weight gain and eventual average carcass weight did not differ between the two feeds. On the other hand, steers in Western Australia grew faster on silage than on hay when the pasture was cut 28 days earlier for the silage and wilted overnight. In this instance, when a barley/lupin mixture at 0, 0.5, 1.0 and 1.5% of liveweight daily was also provided the respective average daily weight gains over 90 days were 0.33, 0.62, and 0.87 kg on hay and 0.81, 1.10, and 1.20 kg on silage.

Overseas results indicate that older cattle usually respond better to silage than younger cattle. Perhaps this effect is associated with the greater breakdown of protein with silage-making and the greater need for protein by young animals.

In America, silage is commonly made from maize. Unlike grass, the digestibility of maize can remain at about 70% of the dry matter from shortly after flowering until the grain is at the dough stage. Supplementing maize silage with 2% urea and minerals has increased the daily weight gain of steers from about 0.6 to about 0.9 kg a day. With young cattle, a supplement of protein meal has given a better result. It has been recommended that silage be made from maize when the grain is at the dough stage. Poor weight gains have been reported with maize silage made from immature plants of high moisture content.

CONCLUSIONS

In fodder conservation the value of the product depends greatly on the maturity of the plants at cutting. The quality of hay is also affected by the speed of drying, the amount of rain falling on the drying herbage, any loss of leaf and any heating or moulding of the hay. In the case of silage, as well as the effect of plant maturity the quality can also be affected by the moisture and sugar contents of the herbage and the degree of compaction in the silo. Wilting can reduce the moisture content and chopping can aid compaction. The silage should be covered with plastic sheeting as soon as possible to stop air movement in the silage and water infiltration.

The relative nutritional value of hay and silage can vary according to the quality of each product.

There can be substantial rotting in large rolls of hay left in the open for prolonged periods.

CHAPTER 8

Mixed Stocking

Both beef cattle and sheep are often kept on the same farm. A pertinent question is whether the two species can complement each other to some degree in pasture utilization. This subject was examined at Canberra and Rutherglen.

RUTHERGLEN EXPERIMENTS

The Rutherglen experiments were conducted on annual pasture. The sheep were autumn lambing ewes producing prime lambs, and the cattle were steers bought at weaning at the start of summer and kept for a year before being slaughtered and replaced. In three experiments, the performance of sheep and cattle grazing together was measured against that of similar sheep and cattle grazing separately, each at several stocking rates.

In the first experiment, where the sheep and cattle grazed together in equal proportions at three stocking rates, the results were as follows:

- At each stocking rate when the sheep and cattle grazed together there was a 10% increase in lamb growth and wool production with no effect on steer growth compared to the sheep and cattle grazing separately.

- The benefit of mixed stocking came from the sheep grazing pasture during late autumn and winter which the cattle were avoiding around cattle dung deposited as long ago as the start of the previous spring.
- On dry pasture the cattle grazed more of the taller, coarser material than the sheep, and only the sheep appeared to eat substantial amounts of clover burr, but these differences had no detectable effect on animal performance.
- An increase in the stocking rate for the sheep and cattle grazing together generally had no more effect on either species than a similar increase in stocking rate for each species grazing separately– indicating strong competition by the cattle against the sheep even on short pasture.
- The only time the sheep had a clear competitive advantage over the cattle was in selecting green shoots from among dry pasture after unseasonable rain.
- The worm burdens were lower in lambs grazing with cattle than with sheep grazing alone.
- A problem with mixed stocking was that the steers required better pasture conditions than the lambs to finish both for market by the end of spring, and the ewes continuing to produce wool even when not gaining weight also favoured the economics of a heavier stocking rate for the sheep.
- Because of this last effect it was more profitable to graze each species of animal separately rather than compromise on stocking rate and graze them together, despite the possible benefits with mixed stocking.
- The most profitable rates of separate stocking were about 8.5 ewes a hectare and about 1.4 steers a hectare, with this cattle-stocking rate providing about the same grazing intensity as 5.6 ewes a hectare.

Findings in later experiments were as follows:

- Having only a small proportion of sheep with the cattle, so that the sheep might exist mainly on the pasture being avoided by the cattle, was of no clear benefit.

- Changing the sheep and cattle over between their separate plots at the opening rains, to reduce the amount of accumulated cattle dung on the cattle pasture during late autumn and winter, was also of no obvious benefit.
- Harrowing the cattle pasture just after the opening rains, to hasten dung decomposition, discouraged the cattle from generally grazing so close to the ground and thereby reduced the overall degree of pasture utilization.

Canberra Results

In the Canberra experiment the sheep were spring-lambing ewes and the cattle were steers bought at the start of spring and kept for 15 months so that the cattle numbers were doubled during spring.

In this experiment, the sheep production also benefited from mixed stocking, but unlike in the Rutherglen experiment there was no appreciable difference between the sheep and cattle in the most profitable stocking level.

In this instance, the similar stocking requirement between the sheep and cattle probably occurred because of the methods of management used. By lambing in spring and doubling the cattle numbers during spring, this would have enabled good pasture utilization while reducing the relative competitiveness between the sheep and cattle at the time of year of scarcest pasture. Also, the ease of finishing the cattle would have been aided by the cattle being kept to a slightly older age and experiencing two periods of spring pasture over 15 months.

General Comments

That sheep are better than cattle at selecting green pasture from among dry pasture has also been found in other experiments, and these results were probably affected by the smaller mouths of sheep than of cattle allowing more precise selection.

At Rutherglen, cattle dung probably persisted in the pasture for so long because the consistency was firm to begin with and for much of the year the sun baked a resistant surface on the pats. In higher-rainfall regions cattle dung generally does not persist as long as it does at Rutherglen, and this would affect the opportunity for improved pasture utilization with mixed stocking.

In Britain, a further disadvantage of pasture harrowing was an increase in the worm burdens in the cattle because the worm larvae were more evenly spread over the paddock rather than being contained to around intact dung pats.

Several sets of results have shown that some worm species do not transmit readily between sheep and cattle, and this can benefit worm control, a subject discussed more fully in the chapter on parasites.

Several sets of results have indicated that where there is a more diverse range of plants, the sheep or cattle may sometimes eat plants that are unacceptable or less acceptable to the other. In Scottish hill farming, having cattle as well as sheep on the hills is reputed to keep the heather plants in a more digestible state.

Conclusions

There can sometimes be opportunities to obtain better pasture utilization and greater animal production from grazing sheep and cattle together rather than separately, but a difference in stocking requirement between the two species of animals may sometimes be a barrier to profitably using this possible benefit.

Sheep and cattle can also often benefit each other in worm control, and sometimes the best way to obtain this benefit may be to change the sheep and cattle over between their separate paddocks after worm treatment.

CHAPTER 9

Growth and Development

Knowing how cattle grow and develop can be useful in deciding management practices.

GROWTH PATTERN

As cattle develop from birth to maturity their daily growth capacity follows a stretched *S* pattern. It increases from birth to about puberty, then stays fairly constant for a substantial period before starting to decline as maturity approaches and fattening increases. At weaning, cattle have generally attained about 80% of their eventual skeleton size but only about 40% of their mature weight.

A relationship that applies for animals in general and appears to influence cattle growth as well is that larger-type animals usually mature more slowly. As a result, large-type animals gain weight faster in absolute terms but more slowly as a proportion of their live weight. In one instance, the daily growth rates between 200 and 400 days of age for large-type Charolais steers and smaller-type Angus were respectively 1.18 and 1.07 kg in absolute terms but 0.30 and 0.35% as a proportion of live weight.

Two effects of this slower maturing of larger-type cattle are a slightly longer time to reach maturity and a few days longer for the gestation period, examples of which are described in the chapter on fertility.

Although the overall growth rate is determined by the potential mature weight of the animals, the actual pattern of growth can vary to some degree between individual animals of similar mature weight. This pattern will be genetically determined, but the specific factors involved can be difficult to identify and could include differences in such things as the hormone levels, appetite, efficiency of digestion, body composition, and capacity to cope with the hazards of the particular environment. In America, cattle within the same breed that grew comparatively well in one environment did not do so on another.

Relative to their height, longer-type cattle usually gain weight faster than blockier-type cattle simply because of a greater body mass at that height. Also, at the same height longer-type cattle have a smaller proportion of their carcasses in the hindquarters and a greater proportion in the loin region, but at the same degree of fatness there is no significant difference between them in the percentage yield of saleable meat.

In a substantial review, it was concluded that a period of feed restriction may not affect the eventual mature size of cattle but will affect the time taken to reach that mature size. In America, the cows in one herd reached their mature size at about 5 years of age, whereas those in another herd in a poorer environment continued to grow very slowly up to about 8 years of age.

In some early work it was claimed that the level of nutrition in early life could affect the eventual muscling of the carcasses, but later work showed that this apparent effect results only from a difference in fatness giving the impression of thicker muscling.

After a period of feed shortage, cattle can exhibit compensatory growth except when the feed restriction occurs during the first few weeks of their lives. In one experiment, calves experiencing severe feed restriction during their first 16 weeks did not show compensatory growth later. On the other hand, in Canada, when steers were fed for a daily live weight gain of 0.23, 0.45, or 0.68 kg during winter, their respective daily gains during spring were 1.14, 1.02, and 0.83 kg. Also, breeding cows that experience weight loss during some part of the year can make a remarkably quick recovery once the pasture conditions improve.

Factors which appear to have contributed to compensatory growth in different degrees in different situations have been increased feed consumption, increased gut fill, an increase in the size of the liver and digestive tract, and better feed conversion efficiency. The better, conversion efficiency probably applied because of the lower fat content in the early weight gain. In one experiment, compensatory growth arose from greater feed consumption and a better conversion ratio of feed to body tissue.

It has also been evident that the total weight gain over a period is likely to depend on the total amount of feed consumed and be little affected by any variation in the feeding pattern. On the other hand, if the same amount of feed is provided over a longer period, a greater proportion of the feed will be devoted to just maintaining the animal.

Body Development

When calves are born, their heads and legs form a relatively large part of the body. Afterwards, the different parts of the body start to develop as they play a more active role in the functioning of the animal. This includes an increase in the belly region as the consumption of solid feed develops. The loin region can be one of the last parts to achieve final proportions.

As cattle grow from birth, the bone and muscle components of the carcass increase at about constant rates, with the rate of increase being faster for muscle than for bone. The rate of fat deposition starts to increase as the animals become older and more mature, so older animals are easier to fatten.

The bone component of cattle can continue to increase very slowly even when they are not gaining weight. It was recorded that cattle reached their full skeleton size at about 55 months of age in the situation where they did not reach their full body size until about 8 years of age.

Growth in the bones ceases when they calcify. This can occur more slowly in steers than in bulls, so steers can continue to grow until they are taller than bulls.

The following results show the difference in the body composition between cattle greatly different in age:

	3 months	45 months
Muscle	68%	49%
Fat	6%	41%
Bone	26%	10%

Fat deposition begins in the body cavity and then spreads to other sites as the animal becomes fatter. Thus, the proportion of fat in each depot changes as the animal becomes fatter.

The fat depots are around the heart, in the body cavity, the subcutaneous fat under the skin, the intermuscular fat between the muscle bundles, and the intramuscular fat within the muscle bundles. This last fat is the last to develop, and it is known as marbling. It can occur to different degrees in different breeds of cattle and is more likely to occur when the cattle are fed grain.

Much of the gain in carcass weight in older animals can be fat. In one experiment, when Hereford steers were slaughtered at 12 or 18 months of age, the respective results were 307 and 495 kg for live weight, 165 and 300 kg for carcass weight, 130 and 180 kg for the weight of saleable meat trimmed to 20% fat content, and 78% and 60% for the proportion of the carcass that was saleable. In another study, an increase in live weight from 505 to 654 kg produced an increase of only 25 kg in the quantity of saleable meat.

Dairy breeds and the large European breeds of cattle tend to have a greater proportion of their total body fat within the body cavity, but the difference has not always been significant.

The ratio of muscle to bone in the carcass can also vary between breed types. It is usually highest in the large European breeds, intermediate in the British breeds, and lowest in the dairy breeds. However, even within the same breed, the degree of muscling can differ greatly between the individual animals, and some animals are so poorly muscled that they are clearly inefficient producers of beef.

The following results show the carcass composition of beef-breed and dairy-breed steers when each was fed at a high or medium feeding level and all were slaughtered at the same live weight:

Breed	Feeding Level	Muscle (%)	Fat (%)	Bone (%)	Muscle – Bone RatioM
Hereford	H	54.5	31.5	11.7	4.7
	M	58	27.7	12.2	4.8
Hereford x Friesian	H	59	26.1	13.5	4.4
	M	62.3	21.6	15.2	4.1

As shown, the beef x dairy breed had a lower proportion of muscle to bone in their carcass, but the percentage of muscle was boosted by a lower fat content.

When an animal loses weight there is a reduction in their muscle, fat, and bone contents, with the proportionate loss being greatest for fat and least for bone. In one instance, when Charolais x Hereford and Hereford steers lost 30% of their live weight from a starting point of 402 and 332 kg respectively, the average results were a loss of about 40% of the carcass weight and 20% of the offal weight. The carcass weight loss consisted of 70% of the initial fat content, 40% of the muscle content, and 10% of the bone weight.

In Victoria, when Angus steers were reared to live weights between 330 and 440 kg and then held at these weights for up to several months before slaughter, the effect during the period of maintaining weight was a slight reduction in the kidney and channel fat and a slight increase in the bone content. Other results have also indicated only small changes in the fat content of the carcass in animals during periods of no weight change.

The amount of energy required to just maintain live weight generally increases by about 73% for each progressive doubling of live weight. Also, it requires more feed energy to produce a kilogram of fat than to produce a kilogram of muscle. Thus, as cattle become heavier and fatter they require more feed for each kilogram of live weight gain. For example, with Friesian and Angus steers fed from 365 to 560 kg live weight, the amount of feed required increased by 2.2 kg, and the daily live weight gain decreased by 0.18 kg for each 100 kg increase in live weight.

Part of the reason it requires more feed to produce a kilogram of fat than a kilogram of muscle is that muscle contains about 75% water whereas fat contains almost no water.

Body tissues is constantly being broken down and renewed with the rate of renewal being greater for muscle tissue than for fat tissue. Also, the rate of regeneration is much faster in the liver and digestive tract than in the other parts of the body, so these parts have relatively high energy expenditures. One effect of this is that when animals are experiencing a long period of severe feed shortage the sizes of the digestive tract and liver shrink in response to the reduced feed consumption, and this substantially reduces the energy expenditure of the animal. Largely because of this effect, the amount of feed required just to maintain weight during prolonged drought feeding can be about 20% lower than normal for that live weight.

Over a given live weight range, larger-type cattle are less mature and so are leaner. As a result, they are depositing less fat and so have a better ratio of liveweight gain to feed consumption. However, when cattle differing in mature size are all reared to the same degree of maturity there is no appreciable difference between them in the overall feed conversion efficiency.

LIVE WEIGHT ASSESSMENT

The value of a carcass depends on the proportions of bone, muscle, and fat it contains and the degree of fat trimming required before the meat can be sold. There are several practices for assessing the live animal which can provide general but not precise results.

One method of assessing the fatness of cattle is to feel the thickness of the tissue cover over the ends of the short ribs of the loin, as is described in Appendix 1. Other indicators of the level of fatness commonly used by farmers are the fullness of the brisket, the general appearance of the animal, and the fullness of the cod or udder. At Hamilton, the short-rib assessments were only about 60 to 80% accurate for assessing the exact thickness of fat cover for the individual animals but were more accurate for assessing the average thickness of fat cover for a group.

It has been claimed that the characteristics of a muscular animal are a square stance, thick muscling on the upper foreleg, a broad back, a belly that sticks out only moderately when viewed from the rear of the animal, a full butt, and a *U* shape rather than a *V* between the hind legs. In one American study, the best indicator of the degree of muscling was the diameter of the upper muscle of the foreleg relative to the live weight. It has been evident from several sets of results that greater fatness can look like thicker muscling.

The thickness of the fat cover and the width of the eye muscle along the back can also be measured using electronic probes. There can be a degree of error in these measurements, and there is also an additional degree of error in using them to estimate the carcass composition. In one study, the composition of the carcasses from young steers fed on pasture could be assessed with about 80% accuracy from using the values for carcass weight and thickness of fat cover, and there was no greater accuracy from including a measurement of the thickness of the eye muscle. With much fatter carcasses the relationships became more complex, probably because of the amount of fat deposited within the muscles.

One set of results indicated that a single measurement of the depth of the eye muscle may provide a poor evaluation of the cross-sectional area of the muscle. For Angus and Hereford steers combined, the average results for accuracy in assessing the area of the cut muscle surface were 44.8% with a single measurement of depth and 83.6% when both width and depth were included.

Sex Effects

Calves of a different sex can start to differ appreciably in their growth rate after about 4 to 7 months of age. Bull calves are usually heavier than steer calves at weaning, which in turn are usually heavier than heifer calves, but the degree of difference has differed substantially between different sets of results. In three New South Wales herds, the average weaning weight was 15.6% lighter for heifer calves and 6.5% lighter for steer calves than that for bull calves, but the results differed between farms and between years on the same farm. With Angus calves at Rutherglen, there was no significant difference in the average weaning weight between steer and heifer calves when the castration was by Burdizzo at the end of calving.

After weaning, several results have indicated that bulls generally grow about 15 to 17% faster than steers, have about a 10 to 13% better ratio of live weight gain to feed consumption, and have less fat in their bodies when assessed at the same age or weight. However, the meat from young bulls is usually more variable in tenderness than that from young steers, and there can be management problems in keeping bulls.

Heifers usually fatten at lighter weights than steers, and at Rutherglen the average difference in the fat thickness of the carcasses at the same weaning weight was 2 mm greater for heifer calves than for steer calves when the overall average thickness of fat cover was 8 mm. In one report, the percentage yield of saleable meat did not differ between steers and heifers at the same degree of fatness, and in another report the yield of saleable meat was 6% lower from heifers than from steers at the same live weight.

The use of growth hormone implants can increase the weight gain of cattle, and the degree of response is usually proportional to the level of nutrition. There are different products available which vary in their relative merit for different situations. Appropriate advice should be sought on their use. In one experiment, when steers were implanted with Zeranol, they consumed 9% more feed, grew 15% faster, and required 5% less feed for a given live weight gain.

Castration of bull calves has sometimes slowed their growth by varying degrees for up to about three months. In one instance there was less stress when the castration was done using a Burdizzo rather than with a knife. In Victoria, castration at ages between 1 and 8 months had no consistent effect on the overall growth rate or on the carcass attributes when slaughter was at 26 months of age. In three overseas experiments in which castration occurred between 3 and 9 months of age and the cattle were slaughtered at 30 months of age or older, the carcass results tended to favour an earlier castration, but no effect was significant. In two of these three experiments earlier castration tended to result in a slightly lighter hide, and in all of them it tended to result in slightly greater fatness.

When castration is by rubber rings it is highly desirable to protect against possible tetanus infection. When a Burdizzo is used, each testicle must be treated separately. This method of castration can be completely reliable if it is done carefully on young calves. If some failures are being encountered, then in future each testicle should be done twice at different places on the testicle cord.

Pushing the testicles up into the body cavity and using an elastrator ring to keep them there creates cryptorchids. These animals have similar growth attributes to bulls, and the relatively high temperature within the body greatly reduces the likelihood of fertile-sperm production. In one set of results, when cryptorchids were created at 4 to 10 months of age, the respective carcass attributes for steers compared to bulls and cryptorchids at a slaughter live weight of 360 kg were 2.93 and 1.50 mm for the thickness of fat cover, 55.6 and 62.4 for the percentage of muscle, 19.5 and 9.3 for the percentage of fat, and 24.9 and 27.5 for the percentage of bone. Slightly tougher meat from the bulls and cryptorchids was probably due to the leaner carcasses. Some of the animals that rendered cryptorchids at 10 months of age still had live semen at 16 months of age.

Cryptorchid characteristics can also be evident in male animals when only one testicle can be removed at castration.

Feeding Effects

Aside from affecting the weight gain, the level of feeding can sometimes also affect the carcass composition. In general, a higher level of nutrition sometimes results in fatter carcasses at a given carcass weight, especially if grain is included in the diet, and a prolonged deficiency of protein can sometimes result in a moderate decrease in the muscle content of the carcass. It has been claimed that cattle have a limited capacity for the rate of muscle formation and that when the feeding level exceeds that requirement, a greater proportion of the feed is devoted to fat production. Some results recorded have been as follows:

a) In one instance, cattle fattened after a long period on drought rations had a slightly-higher-than-usual ratio of fat to muscle in their carcasses.
b) When steers were fed grain and chaff to maintain the same weight gain as those on pasture, the average amount of separable fat in the carcass was 63.7 and 40.8 kg for the respective treatments. Also, cattle fed grain at pasture have been fatter at the same slaughter weight than those fed only pasture.
c) At Hamilton, when steers either grazed pasture after weaning or, in addition, were given hay or hay plus grain during the time of poor pasture conditions, the respective results at a slaughter live weight of 420 kg were 28, 25, and 19 months for the slaughter age, 226, 216, and 226 kg for the carcass weight and 10, 5.5, and 5 mm for the thickness of fat cover.
d) In America, when the feeding practices after weaning were either pasture-grazing, pasture-grazing followed by feedlotting, or feedlotting, the respective carcass results at the same slaughter weight were 59.2, 52.4, and 47.1% for the lean meat content; 5.4,

9.7, and 12.5% for the amount of external fat; and 17.6, 27.7, and 35.4 5% for the total fat content.
e) In one experiment, when feedlot cattle were fed at 100, 90, or 80% of appetite, the average thickness of fat cover at the same carcass weight was 9.9, 8.5, and 7 mm respectively.

DOUBLE MUSCLING

A condition known as double muscling is found in individual animals in several breeds and is most prevalent in the Belgian Blue, Piedmontese, and Parthenais breeds. Belgian Blue cattle have been inbred to stabilize this attribute.

Double muscling arises from a genetic mutation which results in larger muscle cells. The result can be an increase in the muscle content of the animal by about 20–25%. There is little fat in the carcasses. The meat has been reputed to be very tender, but some have disputed this assessment. Adverse effects are delayed maturity, lower fertility, and poorer survival of the calves. These breeds are likely to be used mainly as a source of terminal sires.

Bulls with a modified degree of double muscling are now commonly used in some countries. The degree of supervision that can be applied at calving is likely to affect the merit of using these bulls.

DEHORNING

When mature Brahman cattle, a breed with relatively thick horns, were dehorned in two experiments, the average reduction in growth over the following five months was 11.9 kg on average. When the results from these experiments are combined with those from another experiment, the average amount of carcass trim because of bruising was reduced from 1.34 kg to 0.73 kg. Tipping the horns rather than removing them resulted in an average reduction in liveweight gain of 8.6 kg, with no benefit to the level of bruising.

Sinus problems have been reported for mature cattle dehorned during wet weather.

Calves can be dehorned when young by using a hot iron.

There are regulations in Australia governing castration and dehorning practices relative to cattle age.

Dressing Percentage

The dressing percentage is the ratio of carcass weight to live weight. However, live weight can be measured straight from pasture or after being off pasture for periods of different lengths, together with variable availability of feed and water. Also, the carcass weight can be measured as the hot weight just after slaughter, but a common practice at abattoirs is to deduct 3% from the hot carcass weight to allow for any later loss of moisture and thereby provide an estimate of the chilled carcass weight. In addition, the carcasses can be weighed with the kidneys and kidney fat left in or removed.

In two experiments, the reduction in carcasses weight over the first 24 hours was 0.98 to 1.18%, and in another experiment the reduction in weight over one week was 1.65%.

Regardless of what method of measurement is used, the dressing percentage usually increases with increased animal maturity and fatness. Also, animals consuming more nutritious feed usually have less gut contents, and these gut contents decrease faster during any period of fasting. The gut contents also decrease faster when the animals consume feed of higher moisture content and when they exercise more.

The gut contents can be about 5% of live weight in early life, rising to about 20% at about 200 to 250 kg live weight, and thereafter declining to less than 12%.

CONCLUSIONS

The daily growth capacity of cattle varies with the degree of maturity. Larger-type cattle grow faster on a daily, basis but more slowly relative to their mature size.

The feed requirement for each unit of live weight gain increases as animals become heavier and fatter. There is little difference in the feed conversion efficiency between cattle of different mature size when they are all reared to the same degree of maturity. Differences in the pattern of feeding level can affect the time taken to reach maturity but may not affect mature size.

The carcass composition can be affected by the breed, sex, degree of maturity, and sometimes by the method of feeding. More mature cattle fatten more readily.

Double muscling results in more-muscular carcasses with little fat, but it can adversely affect reproduction.

CHAPTER 10

Milk Production

The milking capacity of cows can affect calf growth and cow fertility and health.

The milk consumption of calves is usually measured by weighing the calves before and after suckling, a kilogram of milk being about one litre. Also, to obtain samples for assessing milk composition the cows are often injected with a hormone to ensure complete milk let-down, because cows can withhold their milk and the composition of the milk varies according to the degree of withdrawal. As part of this effect, the butterfat content can increase towards the end of the withdrawal period.

Colostrum

The milk present in the udder at calving is called colostrum. It is comparatively rich in nutrients, and an especially valuable component is immunoglobulins, which can confer substantial disease resistance to the calf if absorbed through the wall of the small intestine.

The cow's milk contains immunoglobulins for a few days after calving, but the calf's ability to absorb these immunoglobulins declines rapidly after birth and ceases after about 30 to 36 hours. Thus, it is very important that the calf receives colostrum from its dam or another cow as soon as possible after birth. In one set of results, all the calves that suckled within six hours after birth had absorbed adequate levels of immunoglobulins.

SUCKLING PRACTICES

In a British study with 161 Friesian cows mated to Friesian or Sussex bulls, the middle values in the results for the pure-bred and cross-bred calves, respectively, were 81 v 75 minutes to stand, 218 v 130 minutes to first suckle, and 38 v 17% for those that had not suckled after six hours. Calves that had experienced dystocia, were born during the colder months of the year, or were licked less by their mothers were generally slower to stand. After calving, 6% of the cows and 12% of the heifers rejected their calves temporarily or permanently. Many of the heifers seemed to be puzzled by the presence of their calves and kept turning to face them as well as rejecting any effort by the calf to suckle, whereas the cows were more accepting of their calves. Some mature cows were very slow to stand. The size of the teats and the degree of udder suspension greatly affected the time taken to first suckle. Little milk was generally consumed in the first six hours after birth, but the degree of earlier suckling varied between the individual calves.

In beef production, large teats and pendulous udders can be especially common in mature beef x dairy breed cows used in vealer production. With these cows their calves will often suckle only the front teats for a substantial period after birth, unless the cows are hand-milked to reduce tension in the rear teats. If tension builds up in some quarters of the udder milk is resorbed to counter the effect of continuing milk production.

When a calf is consuming milk by suckling or from a bucket a groove forms in the rumen to direct the milk further along the digestive tract. If more milk is consumed at one time than can be digested in the small intestine, then undigested nutrients passing into the large intestine can nourish pathogens, causing scouring. With Angus calves born in early summer at Rutherglen the calves commonly experienced scouring during very hot weather two to three weeks after birth, presumably because of over-consumption of milk at a single suckling. However, in one study, two dairy calves consumed the remarkably high averages of 7.6 and 8.6 kg of milk a day during the second to fourth days after birth without scouring, for reasons that were not identified.

A calf will commonly suckle about eight to nine times a day early in the lactation, declining to about two to three times a day late in the lactation, but the reported suckling frequencies have varied. In an American study, the average number of daily sucklings was 4.6, 4.6, and 3 in successive three monthly periods. Calves receiving less milk have tended to suckle more frequently than the others at the same stage of the lactation.

Some cows, of which Angus is an example, will let other calves suckle at the same time as their own. With these cows, if a calf is orphaned after it is about 6 or 8 weeks old and there is green pasture present, the calf can often continue to grow quite well on stolen milk if left with the herd.

Milk Composition

At Rutherglen, the average composition of milk from Angus cows was 3.6% butterfat, 3.1% protein, and 8.9% solids-not-fat. In some overseas results, the values reported for beef cows have ranged from 3 to 5.8% for butterfat, 3 to 3.9% for protein, and 8.5 to 9.5% for solids-not-fat. Reported average butterfat percentages for dairy cows have been 3.99 for Ayrshires, 3.7 for Friesians, 5.13 for Jerseys, and 4.16 for Brown Swiss. However, within these averages there can be substantial variation in the composition of the milk between individual cows.

In several overseas results, the rate of calf growth has been greatly affected by the quantity of milk consumed, and only in some instances has it been affected by differences in milk composition. In one of these studies, 71.3% of the growth difference between individual calves appeared to be due to the differences in the quantity of milk consumed, 2.7% to the total solids content, 0.5% to the fat content, and the remainder to factors which could not be identified.

Pattern of Milk Production

With well-fed dairy cows, the daily milk production usually increases up to about 4 to 6 weeks after calving, then declines slowly to end on average after about 300 days.

Poorer feeding at any stage of the lactation can result in reduced milk production at that stage, but the normal pattern of production can resume if the feeding level later improves.

Another factor in beef production is how much milk the calves can consume, especially early in the lactation. In several sets of results, the peak in daily milk production has tended to occur later with cows of greater milking capacity, and in one set of results the peak production occurred during weeks one to eight for individual beef-breed cows and during weeks nine to twelve for individual beef x dairy breed cows of greater milking capacity.

Cows rearing twins usually produce more milk. Some of this effect is likely to occur early in the lactation, when single calves can't consume all the milk the cows can produce, but in one instance the average daily milk production at 17 to 22 weeks of age was 9.4 kg with twins and 6.6 kg with singles. The degree of increase in milk production from rearing twins has differed between studies, and in Irish results with Hereford x Friesian cows there was increased milk production with double suckling only in one year in three.

In British results, the growth rate of twins has been 75 to 80% that of singles. Also in some cases but not in others, the milk consumption has been greater with heavier calves at birth, such as bull calves compared to heifer calves. In one study, for each kilogram increase in birth weight the average increase in milk consumption over the first one, two, or three months was 17.2, 32.9, and 41.6 kg respectively, and over eight months it was 200 kg greater for calves 2 kg heavier than average at birth.

Different sequences of pasture conditions arising from different calving seasons can affect the pattern of milk consumption. Results obtained at Rutherglen on this subject are described in the chapter on calving season.

In other instances, the daily milk consumption has declined steadily from its peak a few weeks after calving or has remained fairly constant for several months before declining. With autumn-calving Hereford cows at Hamilton the average daily milk production was 5.4 kg in early October and declined to 1.1 kg in January in a year when the pasture dried off in January. In the following year, the average daily milk production was 5 kg in early October, 3.8 kg in late January, and 0.5 kg in mid-February when the pasture did not dry off until February.

In another experiment at Hamilton with Hereford cows calving in early winter, the average daily milk production was 2.7 kg in July, 2.8 kg in September, 3.7 kg in early November, 2.5 kg in December, and 0.7 kg in March in a year when the pasture dried off in February. In that experiment, a similar pattern of response was recorded with Hereford x Friesian cows, but at a higher level of milk production.

At Rutherglen, partly because of some unusual pasture conditions encountered, the total milk production for the lactation did not differ between calving seasons ranging from early summer to early winter. However, in South Africa the average milk production during lactation declined from 1,485 to 1,029 kg with a later calving season, and the difference in the average weaning weight was 209 compared to 163 kg. There seems to be no readily available results for southern Australia on the effect of a spring calving on milk production.

If cows are in good body condition they can use their body reserves to moderate the effect of poor feed conditions on their milk production. In one study, when the cows were fed a daily ration of 96 or 52 MJ of energy over 150 days the difference in the total milk production was only 1,355 v 1,187 kg, and the difference in the calf weaning weight was only 210 v 199 kg. Also, when Angus cows at Rutherglen calved in good body condition at the start of summer, they maintained sufficient milk production over about five months of dry pasture to support calf growth that did not fall below about 0.4 to 0.5 kg a day, despite the cows losing weight rapidly. On the other hand, with cows that calved in early autumn or early winter, their daily milk production declined sharply after the pasture dried off in late spring or early summer.

The milk production of beef cows may also respond poorly to supplementary feeding. In one instance, the increase in the total milk production from providing the same amount of supplement over 105 days was 300 kg for dairy cows but only 37.5 kg for beef cows.

Milking Capacity

The milking capacity of cows varies between breeds and between individuals in the same breed.

In one study with Angus milked by machine, the total milk production during the lactation ranged from 467 to 3,066 kg for the individual cows.

With autumn-calving cows at Hamilton, the average daily milk production in midwinter over three stocking rates was up to 5.9 kg with Hereford cows, 7.1 kg with Hereford x Angus, and 9.3 kg with Hereford x Friesian. In American results, the average daily milk production was 5 kg for Herefords, 6 kg for Angus, 4.5 kg for Limousin, 5 kg for Charolais, and 11 kg for Simmental. In several Australian results, the increase in the calf weaning weight from the greater milk production of beef x dairy breed cows compared to beef-breed cows has been about 10% on average.

When two breeds differing in milking capacity are crossed, the milking capacity of the progeny can be greater than the average for the two parent breeds. In one instance, the milking capacity of the progeny was similar to that of the better-milking parent breed.

Rearing Effects

If heifers become fat before they reach puberty, then fat deposits in the udder can suppress the development of milk-producing tissue, leading to reduced milking capacity later. One set of results indicated that the earlier the fatness develops the more likely it is to affect the later milking capacity, and other results have indicated that differences in the level of nutrition after puberty appear to have little further effect. Another set of results indicated that if heifers gain more than about 1 kg a day in live weight before they reach puberty there may be an effect on their eventual milking capacity.

In a study of 1,626 heifers from seven Hereford herds and one Shorthorn herd in adjoining parts of South Australia and Victoria, a trend recorded was that for each 10 kg increase in the weaning weight of heifer calves in the range of 170 to 250 kg, there was generally a 5 kg reduction in the eventual weaning weight of their first calves.

In an overseas experiment, unrestricted feeding of heifers from 7 to 12 months of age resulted in an average reduction of 36 kg in the weaning weight of their first calves. Also, when dairy heifers were unrestricted or restricted in their weight gain from 7 to 12 months of age, the average results were 345 v 438 days for the time to reach puberty, 12.6 v 13.9 kg for the average daily milk production during the first lactation, and 245 v 268 days for the length of the lactation.

In American herds, a pattern was observed of an alternating moderate rise and fall in the average weaning weight between successive generations, indicating that the milk production in one generation affected that in the next. Also, in a large American experiment it was found that heifer calves that received more milk during suckling because of the age of their dams generally had a lower eventual milking capacity. For example, heifers that had suckled 2-year-old dams on average produced 0.63 kg more milk a day and weaned progeny 9 kg heavier than those that had suckled 5-year-old dams.

While there can be benefits to milk production from restricting the feed consumption before puberty, a set of Australian results indicated that with heifers to calve at 2 years of age the emphasis should be on adequate development for conception and rebreeding, as described in the chapter on fertility.

Cow Age

In the various reported results, milk production and weaning weight have generally increased up to cow ages of 5 to 8 years and then varied from a slight increase to a slight decrease with older cows. In an American study for cows aged less than 6 years, 6 to 9 years, and over 9 years, the respective average results were 1,117 kg, 1,580 kg, and 1,418 kg for milk production and 172, 207, and 202 for calf weaning weight, and in another American study the average weaning weights for calves from cows aged 3, 4, 5–7, 8–11, and over 12 years of age were 219, 225, 229, 220, and 202 kg respectively.

In a four-year survey of four New South Wales farms, the average weaning weight of calves from heifers or cows aged 2, 3, 4, and over 8 years differed from that of calves from cows 5 to 8 years old by –15, –8.56, –4.5, and +1% respectively. However, the pattern of result differed between years. This included values which ranged between years from –10.3 to –26.9% for 2-year-old heifers and from +2.74 to –3.7% for cows over 8 years old.

In American results, the level of milk production at 3 and 4 years of age was greater if the cows had first calved at 2 years of age rather than 3, indicating that both age and the number of times calved was affecting the development of the milking capacity in these early years.

The exact effect of cow age on weaning weight on different farms is also likely to depend on how heavily young cows are culled for poor production. It may also depend on how differences in the pasture conditions enable older cows to maintain their nutritional status and enable the calves to compensate for lower milk consumption by consuming more pasture. In American results, the effect of cow age on the calf weaning weight was greater in drier regions with poorer pasture conditions.

Even in situations where the weaning weight may decline slightly in older cows, American results have shown that retaining older cows in the herd as long as they are functioning well can increase the number of progeny for sale and improve herd profitability. Older cows can breed more reliably and calve more easily than younger cows, especially compared to heifers.

Calf Response

Calves can partly compensate for lower milk production by starting to consume pasture earlier. Earlier pasture consumption results in earlier development of the rumen. By 4 weeks of age calves can be consuming measurable amounts of pasture, and by 8 weeks of age they can be digesting pasture well. At this age they can be weaned successfully on to good-quality pasture, but this is about a minimum treatment.

In different studies, the differences in milk consumption have accounted for 20 to 70% of the difference between individual calves in their growth to weaning. In one study, the difference in milk production accounted for 68% of the differences in the weaning weight in one year and 50% in another. In South African results, when the average levels of daily milk consumption were in the ranges less than 4.5, 4.5 to 6.8, and over 6.8 kg a day, the average weaning weights were 163, 196, and 230 kg respectively. Also, in Western Australia when the average daily milk production in different breeds was 3.6 and 4.9 kg, the resulting average weaning weights were 223 and 286 kg respectively.

As calves grow, their milk production becomes a smaller proportion of their total diet, so they become more dependent on the pasture conditions for their nutrition. Some overseas results have indicated that milk consumption may have little effect on calf growth after about 5 to 7 months of age. However, with autumn-calving Hereford cows at Hamilton, where the pasture usually dries off in January or February, weaning before the pasture dried off resulted in a lower yearling weight. The average yearling weights from weaning at 4, 6, 8, or 10 months of age were 210, 224, 251, and 262 kg respectively. An exception was that calves weaned at 8 months of age on to the especially nutritious pasture regrowth that followed haymaking had a similar yearling weight to those weaned at 10 months of age.

In a British experiment, when Angus and Charolais x Angus calves were weaned on turning out to pasture at 7 months of age rather than being left with their dams until 11 months of age, the earlier weaning resulted in a 30–35% reduction in growth rate during the last four months. Also, with autumn-born calves in America, delaying the weaning date from 3 May until 27 June resulted in an increase in the weaning weight of 8.9 kg, but 4 kg of this advantage was lost during the first seven weeks after weaning.

At Hamilton, earlier weaning of the calves from autumn-calving cows had no consistent effect on the cow weight change, but under very poor pasture conditions there may be a benefit to cow weight and fertility from earlier weaning.

In four sets of results, much of the difference in the weaning weight from a difference in milk consumption persisted for the following year, and at Rutherglen, differences in the weaning weight from a difference in the calving season were not reduced during the following year. However, in an American set of results the immediate post-weaning growth was inversely proportional to the pre-weaning growth, with the range in the daily post-weaning growth rate being 0.943 to 0.983 kg a head daily. Overall, therefore, it seems that the degree of persistence of any weaning weight benefit from greater milk consumption may vary between different situations for reasons that are not clear.

Cow Fertility and Health

Greater milk production places greater nutritional demand on the cows, and if this is not met there can be poorer fertility, especially in the rebreeding of 2-year-old heifers. This subject is covered more fully in the chapter on fertility. There is also a greater likelihood of metabolic disease with greater milk production because milk on average contains 1.2 g of calcium and 2.4 g of magnesium per litre. Milk fever can be common in beef x dairy breed cows, and grass tetany can be common in lactating cows grazing pasture with little clover or when grazing cereal crops.

Productive Life

Because of greater udder tension together with sometimes poorer rebreeding and greater likelihood of metabolic disease, it could be expected that beef x dairy breed cows could have a shorter productive life than beef-breed cows. This was not the case in one American study but in another the crossbred heifers rebred poorly. Also, in Scotland the average results for Shorthorn x Galloway and Hereford x Friesian cows over ten years were 90 v 83% for the pregnancy level, 17 v 27 days from the start of mating to conception, 364 v 374 days for the interval between successive calving, and a higher culling rate with the beef x dairy breed cows than with the beef breed cows.

Overall Effects

Results obtained at Hamilton show the effects of the stocking rate and milking capacity on cow productivity and profitability. The cows were Hereford (H), Hereford x Angus (H x A), and Hereford x Friesian (H x F), and they were mated to Hereford bulls for an autumn calving. The daily milk consumption was measured in winter. The average treatment results were as follows:

	Cow Breed	Cows/Hectare		
		0.9	1.3	1.7
Daily Milk	H	5.9	4.4	3.4
Intake in Winter	H x A	7.1	6.3	4.1
(kg/calf)	H x F	9.3	8.3	3.9
Weaning	H	283	247	225
Weight (kg)	H x A	305	287	269
	H x F	330	301	264
Pregnancy	H	100	94	84
Rate (%)	H x A	100	97	95
	H x F	100	84	62
Gross Margin	H	151	138	88

| ($/ha) | H x A | 167 | 196 | 188 |
| H x F | 196 | 175 | 26 | |

These results indicate three effects:

1) Cows of low milking capacity can be relatively unproductive and relatively unprofitable when assessed on calf weaning weight.
2) For any breed of cow, the most profitable stocking rate is likely to be the heaviest at which a high level of fertility can be maintained.
3) There may be no benefit from using cows of greater milking capacity than that found in good milking strains of beef-breed cows, because of the better pasture conditions required to support such an increase in milk production.

Other results have also shown that the most profitable stocking rate for beef cows is likely to be the heaviest at which a high level of fertility can be maintained. Moreover, in one set of American results, beef x dairy breed cows proved to be relatively unprofitable because of their higher feed requirement.

When the Hereford x Angus cows at Hamilton were stocked at their most profitable level the calves were of vealer standard at weaning. Results obtained at Rutherglen support this evidence that beef-breed cows of good milking capacity can produce calves of wealer standard at weaning if the cows are stocked at their most profitable level and calve no later than early autumn.

Another effect apparent in the Hamilton results is the benefit of hybrid vigour. This effect is evident at the middle and heaviest stocking rates as the better fertility of the Hereford x Angus cows than of the Herefords, despite the greater milk production of the crossbreds. The degree of benefit from cross-breeding is usually greater with more-challenging circumstances.

Conclusions

Calves should receive colostrum from their own or another dam within a few hours of birth. How readily this occurs depends on dam temperament and teat and udder dimensions. Later, calves may suckle other dams as well as their own.

Allowing heifers to become fat before puberty can reduce their later milking capacity. Milk production can change continuously upward or downwards according to changes in the pasture conditions. Also it can fall more slowly on dry pasture at the start of the lactation than on dry pasture later in the lactation. Calves should generally not be weaned until the pasture dries off.

Cows can use their body reserves to support milk production and may respond poorly to supplementary feeding. Beyond a cow age of about nine years old, the calf weaning weight may vary from a slight increase to a modest decrease, probably depending on the pasture conditions. There can be financial benefit from retaining older cows that continue to perform well.

After they are a few week old calves may increase their pasture consumption to partly compensate for lower milk consumption. In some instances, heavier calves at birth may consume more milk.

Overall, it seems that the most profitable practice is likely to be to use beef-breed cows of good milking capacity and stock them at the highest level at which good fertility can be maintained.

CHAPTER 11

Calving Season

The calving season used can affect the animal performance in various ways.

Calving seasons were examined in detail in Rutherglen and more generally elsewhere.

RUTHERGLEN RESULTS

The Rutherglen study was conducted with Angus cattle on annual pasture. The calving seasons used were early summer (ES), early autumn (EA), and early winter (EW), and from nine-week mating periods they were timed to begin on 20 November, 1 March, and 1 June respectively. The ES calving began ten days before the official start of summer, to reduce the number of calves likely to be born during very hot weather.

The calves were weaned at the end of October, January, and April for the respective calving seasons at an average age of 10.5 months. Weaning at this age enabled each treatment to show its full potential. The only supplementary feeding was licks in two winters with little clover in the pasture to protect against grass tetany.

The experiment continued until three crops of calves had been reared for each calving season and a fourth born. Half the calves were slaughtered at weaning and the others were kept for a year afterwards to examine the effect of calving season on post-weaning growth. The average period of pasture growth at this location is from late April until early November, and there is usually only dry pasture for the rest of the year. During the experiment, the winter pasture conditions were average, poor, and very poor in successive years, and in two years there was the unusual situation of unseasonable green pasture up to 10 cm tall from late February in one year and up to 5 cm from early March in the following year. Because of the timing or the different stages, the last lot of ES and EA calves experienced the first period of unseasonable green pasture when they were very young or just being born, and the last two lots of EW calves each experienced a period of unseasonable green pasture in the last few months before weaning at the end of April. This variation in the pasture conditions added to the response information obtained.

The results obtained and how they compare with other results from South Australia are described below.

a) *Weaning weight.* The average weaning weight for each calving season for each year was as follows:

Calving Season	Year 1	Year 2	Year 3
Early Summer (ES)	292	292	220
Early Autumn (EA)	292	291	298
Early Winter (EW)	296	295	279

As shown, the only marked difference in the average weaning weight was at EW, where the result ranged from well below to similar to that at the other two calving seasons and was proportional to the amount of unseasonable green pasture the EW experienced in their last few months before weaning at the end of April.

Results from other Australian locations have also shown a reduction in weaning weight with a later calving season, but the degree of effect has varied and has probably been affected by such things as differences in the calving and weaning times relative to the period of pasture growth.

Australian results have been as follows:

Location	Calving Times	Average Weaning Weight (kg)
Holbrook (NSW)	Mid June v Late August	216 v 163
Mansfield (Vic)	June v August/September	243 v 224
Meredith (Vic)	April v September	228 v 202
Hamilton (Vic)	May/June v August/September	251 v 205
Hamilton (Vic)	May/June v August/September	241 v 220

American results have also shown that the calves born early in the calving period are usually the heaviest at weaning.

b) *Milk consumption.* At ES, the average daily milk consumption declined slowly from 6.5 to 4.9 kg on dry pasture and then increased to about the initial level with the start of pasture growth before declining again slowly. At EA and EW, the average value stayed at about 6.5 kg until the pasture dried off at the end of spring and then decreased rapidly to reach a negligible level about a month later at EA and about two months later at EW. However, despite these differences in the pattern of milk consumption, total milk consumption for the lactation did not differ significantly with calving season.

During the periods of unseasonable pasture growth, the milk production increased substantially only at EW, where it occurred during the last three months before weaning.

Results from other locations have also shown increased milk production with improved pasture conditions in mid lactation and later a rapid decline in milk production when the pasture dried off.

c) *Calf growth.* The capacity of calves to grow increases up to about puberty, and as suckling calves become older and heavier they become less dependent on milk consumption and more dependent on the pasture conditions for their continuing growth. These relationships influenced the weaning results in Rutherglen in the following ways:

1) In the first two to three months after calving, calf growth was affected little or not at all by the differences in the pasture conditions, because of adequate milk consumption relative to light live weight.
2) By three months after calving, milk consumption in ES was declining, but with these light calves it was still adequate to maintain an average calf growth rate that did not decline below. 0.4 to 0.5 kg a day.
3) With the start of pasture growth, the level of milk production at ES and EA increased to the initial levels and then started to decline slowly at ES and stayed about constant at EA.
4) During spring, the average daily calf growth was 1.2, 1.3, and 1.1 kg at ES, EA, and EW respectively, because the growth capacity increased with age, but the growth at ES was restricted by reduced milk consumption
5) When the pasture dried off in early November, milk production declined rapidly to reach negligible levels by the end of December at EA and end of January at EW.
6) During the two periods of unseasonable pasture growth, the only substantial responses were in EW because it was only there that the calves were old enough and the stage of lactation advanced enough for much of a response to be obtained.

Overall therefore, in a year of about average pasture growth an earlier calving season resulted in better use of milk production to support calf growth on dry pasture and developed in the calves a greater capacity to grow during the period of pasture growth.

d) *Calf birth weight.* The level of dam nutrition during the last third of pregnancy can appreciably affect calf birth weight, and calf birth weight can greatly affect the level of dystocia.

In the Rutherglen experiment, the average calf birth weight was 2 kg lower on average at ES than at EW and at EA it varied between these two levels in proportion to how much green pasture these cows experienced just before calving.

With 27 cows at each calving season each year the total number of difficult births or stillborn calves over four years was 3, 6, and 10 for the successive calving seasons. Therefore, having the cows calve on dry pasture can sometimes result in lighter birth weights and less dystocia. A low level of dystocia from calving on dry pasture was also recorded in later experiments at Rutherglen.

At Kangaroo Island there also tended to be more calving difficulty with a later calving time in the period from autumn to winter and in Western Victoria calf birth weight has sometimes tended to increase from autumn into spring.

The relatively light birth weight at ES in the Rutherglen experiment occurred despite these cows just having experienced a period of spring pasture. This result may have been due to these cows being relatively fat at calving because fat cows in Britain tended gave birth to lighter calves.

e) *Carcass values.* The ES and EA calves were all of vealer standard, whereas the EW calves were of this standard only in the year with the greatest amount of unseasonable green pasture. Also, despite a similar weaning weight the average calf carcass weight was 7% greater at ES than at EA, reflecting the higher dressing percentage that commonly applies with cattle slaughtered off more nutritious feed.

On a farm in central Victoria, calving in June compared to spring resulted in progeny that were 12.5% heavier at weaning, 15% heavier at 88 weeks of age, and required less supplementary feeding. Also, on a farm at Rutherglen with purchased weaners, those born in autumn could be slaughtered at the end of their second spring whereas those born in spring had to be kept until the end of their third spring but were heavier as slaughter.

f) *Cow weight.* The cows that calved in early summer commonly lost weight at about 1 kg or more a day on dry pasture, and generally for each three-month advance in the calving season from early winter the average minimum cow weight for the year was about 20 kg lower.

The minimum weight reached was not a problem until the year when the pasture growth in winter was very poor. Then the cows at the two heavier stocking rates reached critical weights, and the degree of stress differed little with calving season. Similarly, at Canberra a later calving season did not reduce the stress level in the cows in a year of scarce winter pasture. Therefore, at least in some instances a later calving season may not provide greater safety in such years.

g) *Fertility.* Cow weight at calving and weight change afterwards can affect the fertility level. In the Rutherglen experiment the calving rate did not differ with calving season, and overall was 88%. Also, despite the stocking rate ranging from 0.6 to 1 cow per hectare, the fertility level was generally the same at each stocking rate.

At Hamilton and Canberra, the fertility level also did not vary with calving season, but it differed greatly with stocking rate. Other results have also indicated that the fertility level is likely to vary with stocking rate, and why this did not happen at Rutherglen is not clear. The heaviest stocking rate at Rutherglen was clearly about the heaviest practical without supplementary feeding.

h) *Disease.* The only clear difference in the disease incidence with calving season was scouring in many of the ES calves during the first few weeks after birth. Laboratory tests showed no unusual cause for this effect, so it was probably just due to overconsumption of milk during the hot weather of summer when the calves were very young. This condition was generally well controlled by dosing the worst affected calves with anti-scouring tablets and placing the few extreme cases on electrolyte-replacer for three days. However, only on some farms near Rutherglen has scouring been a problem with an early calving, and the reason for this difference was not clear.

Also, at Rutherglen having the ES cows fat at calving and then losing weight rapidly afterwards could have made them prone to suffering from fatty liver syndrome, which is associated with rapid fat loss, but this did not occur.

i) *Post-weaning growth.* In the twelve months after weaning, growth did not differ consistently with calving season, so any effect of using a different calving season on weaning weight was retained. In overseas results, however, weaning weight differences have sometimes declined progressively.

Conclusions

Overall, it is evident that calving about two to three months before the start of the main period of pasture growth can result in lighter birth weights, less dystocia, better use of milk production to promote calf growth on dry pasture, and the development in the calf of a relatively good capacity to grow on green pasture, with no adverse effects on other factors.

In Victoria, calving in early summer is probably likely to have most benefit in regions of annual pasture by enabling the calves to be sold as vealers before beef prices fall to their lowest point of the year with the increased selling of cattle from the main beef cattle regions of higher rainfall.

The cows used at Rutherglen probably had an optimum level of milk production for use with an early summer calving, and a greater or lower level of milk production would probably have required uneconomic use of supplementary feeding to support either cow fertility or calf growth.

Early-summer-calving cows at mating and continuing to lose weight.

Early-summer-born calves at weaning at end of October.

Early-autumn-born calves in late October.

Early-winter-born calves in late October.

Early- summer-born calves at 22 months old.

More even grazing in September when cattle with sheep than alone.

CHAPTER 12

Fertility

The fertility level achieved has been shown to be one of the most important factors in affecting herd productivity and profitability.

SURVEY RESULTS

In a survey of 18 New South Wales properties in 1968, the average pregnancy results were 84% for heifers, 87% for cows rearing a calf at mating, and 84% for those not rearing a calf at mating. The marking percentage was lower than the fertility percentage by 17.5 and 3.5% for the heifers and cows respectively, reflecting the calf losses from various causes.

In a survey of 13 Victorian herds in 1969 and 1970 where many of the heifers first calved at 2 years of age, the overall average values were 88.7% for the calving rate, 7.85% for the calf losses between birth and weaning, 1.5% for the twinning rate, and 361 days for the calving interval. For cows aged 3, 4, 7–8, 9–11, and 12 years and over, the calving rates were 81.5, 90.4, 95.9, 91.9, and 85.7% respectively. The relatively low value at 3 years of age reflects the difficulty sometimes experienced in getting heifers back in calf after calving at 2 years of age.

HEIFER MATING

Several sets of results have shown that the lifetime production of beef cows can often be increased by having them first calve at 2 years of age rather than three, but the degree of benefit has varied. Some of the reported results have been as follows:

a) In three experiments, the improvement in lifetime production from the earlier calving ranged from 0 to 1.19 calves per cow in the different experiments.
b) In an American study which continued until the 14th calving with cows that first calved at 2 or 3 years of age, the results for the respective ages at first calving were as follows: cows remaining at end of the study 39 and 37%, assistance at first calving 47 and 2%, total weight of calves weaned 533 and 482, and average weaning weight 216 and 220 kg.
c) In a Canadian experiment, the degree of benefit from an earlier first calving had declined to only 40 kg of calf per cow by 8 years of age, the suggested cause being an adverse effect on the milking capacity from the more generous feeding required to have the heifers calve at 2 years of age, an effect explained in the chapter on milk production.

Clearly, with favourable management there can be opportunity for greater lifetime production from having heifers first calve at 2 years of age rather than 3, but the merit of the practice will depend largely on a) the cost of any extra supplementary feeding required, b) how well dystocia can be controlled at the first calving, and c) how well the heifers will rebreed after their first calving.

The indication that a heifer is ready to breed is a willingness to stand and be served. Beforehand, there can be one or more periods of silent heat in which the heifer ovulates (sheds an egg from its ovaries) but does come on heat.

On average, puberty is commonly reached at about 52% of the mature weight in beef breeds and at about 45% in dairy breeds, but the level of feeding can affect the result, and crossbreds can sometimes reach puberty a little earlier than purebreds.

Better feeding can result in puberty being reached at a younger age but sometimes at a heavier weight. In one instance, when groups of heifers were fed to gain weight at 0.3 or 0.7 kg a day from 7 to 12 months of age and then run together, the respective average results for puberty being reached were 14.1 and 12.6 months of age and 239 and 259 kg in live weight. The greater weight at puberty with more generous feeding may have resulted in greater body fatness relative to the degree of development of the reproductive organs.

In partial contrast to this last result, in one instance in Australia better feeding of Hereford calves resulted in puberty being reached at a younger age but with little difference in weight.

There can be a breed difference in when puberty is reached. In one instance, the average results at puberty for Hereford, Angus, and Shorthorn heifers respectively were 413, 337, and 318 days of age and 306, 251, and 243 kg live weight. In another set of results, the average time (and range between individual heifers) at which puberty was reached was 358 (346–370) days of age for Angus and 390 (372–408) for Hereford. Larger-type cattle generally reach puberty at an older age, and it is likely that a difference in mature size affected the above results for breed differences.

American results indicate that heifers generally need to reach 55 to 65% of their mature weight at first mating to be highly fertile.

At Hamilton and on two South Australian farms, an appropriate target mating weight for Hereford heifers to calf at 2 years of age proved to be about 270 kg, but this value would be likely to change over time with any change in the mature size of the animals. In the South Australian results, the pregnancy level increased by an average of 7% for each 10 kg increase in the live weight at mating in the range of 175 to 265 kg, and did not change further beyond this point.

To have heifers calve at 2 years of age usually requires better feeding, and this affects the economics of the practice. With autumn-born calves at Hamilton, the level of supplementary feeding required from late January to have the heifers heavy enough for mating in early June was good-quality lucerne hay (18.9% crude protein) provided to appetite, or a supplement of good-quality oats (11.9% protein) provided at a daily level of 1.6% of live weight. However, grain-feeding of cattle can result in increased fat deposition, so it may be better to avoid the use of grain before puberty, or at least restrict the level at which it is provided.

When Hereford heifers in South Australia to be mated at 15 months of age were weaned at 2 months of age and given different combinations of feeding levels between 3 and 14 months of age, this affected their later milk production. Measured over the first three calf crops, the effect of a higher level of feeding during rearing generally resulted in lower milk production and a lower weaning weight, but earlier puberty, better fertility, and a heavier total weight of calf weaned per cow. The range in average weaning weights was 242 to 210 kg, and the range in total weight of calf weaned per cow was 465 to 573 kg. Milk production was generally affected to a greater degree with an earlier provision of a higher feeding level.

This last set of results indicates that in rearing heifers to mate at 15 months of age the emphasis should be on adequate feeding to achieve early puberty and sufficient development by 2 years of age for successful rebreeding. With heifers to be mated at a later age there is likely to be more opportunity to improve milking capacity by restricting feeding before puberty.

Heifers calving at 2 years of age can be slow to return to heat, so it is commonly recommended that their mating should begin up to four weeks before the main herd. In one experiment, the average time taken to return to heat after calving for 2-, 3-, and 4-year-old Angus x Hereford cows was 88.4, 62.8 and 57.4 days respectively. Also, in New Zealand the average interval from calving to first heat was 113 days for heifers and 84 to 95 days for cows of different ages. In this last instance, the corresponding values for the interval to the next calving were 405 days and 373 to 385 days.

In Western Australia, the average time taken to return to heat after calving was 60 days for 2-year-old heifers and 51 days for cows.

In Britain, a regular 12-month breeding cycle was achieved after the second calving with heifers first calving at 3 years of age, but not until after the fifth calving with those first calving at 2 years of age. The management practices probably affected these results.

It has also been found that if the heifers are heavy enough to conceive, those that conceive relatively early at their first mating are likely to be relatively fertile over their whole life and those that fail to conceive are likely to continue to breed relatively poorly. Therefore, it has been recommended that only heifers calving early in the calving period should be retained as herd replacements. Coarse, masculine-looking heifers can also be irregular breeders.

One claim was that heifers can be tested for pregnancy six weeks after service. Another claim was that some operators may be able to pregnancy test at 35 to 45 days but others may not be able do it reliably until about 50 to 60 days. This method of pregnancy testing can result in about 1 to 3% of abortions in some instances. Moreover, blood testing at about 90 days can detect pregnancy with an accuracy of about 98 to 99%, and an ultrasound reading at 28 days can additionally identify the sex of the calf and the presence of twins.

Return to Heat

Cows commonly first ovulate about 13 to 40 days after calving, and the uterus has generally returned to normal by about 35 to 40 days after calving, times that can be independent of the nutritional status of the animals. One claim was that it is not possible to become pregnant during the first two to three weeks after calving, and that there is a slight possibility of becoming pregnant during the following two to three weeks before regular breeding cycles resume. Especially if the feeding has been poor, the cows can experience periods of silent heat in which they are ovulating but not showing any external signs of doing so.

In one instance, the time taken to return to heat after calving varied from 46 to 168 days for individual cows, and in another instance the proportion of cows showing heat at 40, 60, and 80 days after calving was 30, 72, and 90% respectively. Other results have shown considerable variation.

Several results have shown that the body condition of heifers and cows at calving can be most important in affecting the time taken to return to heat and become pregnant, and that if the heifers or cows are too lean at calving they may require a period of weight gain before they start exhibiting heat periods. Some results reported were as follows:

- In America, when body condition was assessed on a scale of 1 to 9 (emaciated to obese), heifers which calved in condition score 4 did not return to heat on average until 130 days after calving, and the conception rate was only 16%, whereas the corresponding figures for heifers calving in body condition 5 or greater were 60 days and 75%.
- The results in a three-year experiment indicated that if mature cows calve in body condition score no less than 5 (on a scale of 1 to 9), the pattern of weight change during lactation is unlikely to affect rebreeding performance.
- In one experiment, better feeding after calving resulted in a faster return to service only in the cows calving in poor body condition.
- In America, when cows calving in condition score of 4 (on a scale of 1 to 9) were fed at four different levels after calving, with the lowest level resulting in a daily weight loss of 0.2 kg a day, the eventual pregnancy level was 68%, with the poorest feeding level and 85 to 98% for the other levels. The poorest level of feeding also reduced the weight gain of the suckling calves.
- At Rutherglen, when Angus cows calved mainly in condition score 3.5 to 4 (on a scale of 0 to 5) on dry pasture at the start of summer and lost weight at 1 kg a day or more for several months afterwards, their rebreeding performance was as good as that of similar cows calving later in the year under more-favourable pasture conditions.

- In America, when 2-year-old heifers were of condition score 4, 5, 6, and 7 (on a scale of 1 to 9) just before calving, the respective average results were 92, 82, 74, and 76 for the days to pregnancy and 64.9, 71.4, 87, and 90.7% for the pregnancy level.
- In a British study, the average calving interval for cows was 364 days for those calving in condition score 2.5 to 3 (on a scale of 0 to 5) compared to 418 days for those calving in condition score 2.
- In a review of the results of several experiments, a conclusion was that if cows calve in good body condition the results on average are likely to be about 60% of the heifers and about 90% of the cows returning to heat by 60 days after calving.
- Heifers calving at 2 years of age are immature, and several sets of results have indicated that they should be fed to gain live weight at least slowly after calving.

American recommendations, based on a substantial number of results, are that when assessed on a scale of 1 to 9, the minimum condition scores at calving should be 6 for heifers and 5 for mature cows. A Scottish recommendation is that on a scale of 0 to 5, cows are likely to rebreed satisfactorily in body condition score of about 2.25 if they are about to be turned out to graze on spring pasture and in body condition score of about 3.5 if they are about to experience poorer feeding from being taken inside for the winter. Details on these two methods of condition scoring are given in Appendix 1.

Another consideration in Australia is to have the cows in sufficient body condition at calving so that there is only an occasional need for supplementary feeding if the cows regularly experience a period of poor pasture conditions after calving.

The desirable condition score at calving may differ to some degree between breeds. Large European beef breeds, dairy breeds, and beef x dairy breed cattle generally have a greater proportion of their fat within the body cavity than British beef breeds, so they may have a lower condition score at a given level of total body fatness. In one instance, the condition score at the same degree of body fatness was just over 3 for British breed cows but only 2.4 for Hereford x Friesian cows. With beef x dairy breed cows an allowance may also have to be made for the likely effect of greater milk production on the return to heat.

Dystocia and retained afterbirth have also been implicated in some instances in a slower return to service after calving, and in one study the pregnancy results for cows that did or did not experience dystocia were respectively 54 and 69% after 45 days of mating, and 69 and 85% after 105 days. The results of three other studies indicated that with dairy cows the occurrence of dystocia, retained afterbirths, milk fever, and twin calves tended to delay the return to heat by about five to eight days.

In one experiment, exposing heifers to vasectomized bulls from six weeks before the start of mating resulted in the average calving date being 5.5 days earlier and the length of the calving period being 2 weeks shorter. However, a similar result has not occurred consistently in other experiments.

CALF SUCKLING/MILK PRODUCTION

Cows suckled by calves can be slower to return to heat after calving than those milked by machine. This effect is thought to operate through an influence on the hormone levels of the cows. Some of the reported results were as follows:

- The average time taken to return to heat was 46 days for cows milked twice a day by machine compared to 72 days for those nursing a calf.
- In five trials, the number of days to first heat was 54 to 93 days for cows being suckled compared to 18 to 41 days for those whose calf had died.
- Adding a foster calf to Hereford x Friesian cows resulted in a reduction from 71% to 43% in the proportion of cows showing heat by 90 days after calving, despite no difference in the weight change between the different cows as a result of supplementary feeding.

- When cows suckled 2, 3, or 4 calves, the effect on fertility was affected by the total time spent suckling and not by the number of calves suckling each cow.
- In one study in which the hormone levels in the cow were used to measure the time of ovulation, the period from calving to ovulation was on average 29 days when the calf was removed at birth, 31 days when the cows were milked by machine twice a day, and 40 days when the calves were left with the cows.

Removing the calves for two days three weeks before the start of mating was sometimes of some benefit in shortening the time to return to heat in American experiments but not in two Australian experiments. In another instance, allowing the calves to suckle only once a day from 21 to 30 days after birth seemed to result in a quicker return to service. It was claimed that the practice is likely to be beneficial only with very lean cows.

One set of results indicated that if cows stop coming on heat because of a loss of weight, they may have to return to a heavier weight before the heat periods resume.

Conception

There can be poor conception from mating cows at their first heat after calving. It was suggested that this may be due to a non-sterile environment in the uterus at that time. In one instance, the conception rate from mating at the first heat was 20% compared to 78 to 82% from mating at later heat periods.

The interval between successive heat periods is generally about 21 days in cows and 20 days in heifers. In two experiments, the interval between heat periods after calving was 16 to 17 days between the first two heat periods and 21 days between the second and third heat periods. Sometimes irregular heat periods can indicate some abnormality. Each heat period usually lasts about 9 to 20 hours in cows but is often shorter in heifers. In one set of results for cows, the average length of the heat period was 16.9 hours with a range of about 7 to 26 hours.

One problem that can be associated with more-frequent heat periods than normal can be cystic ovaries, something that can be most common in cows of about 5 or 6 years of age and older. The affected cows roar like a bull. Sometimes the condition can be treated, but this may not be economic.

It has been estimated that ovulation probably occurs about 12–15 hours after the end of the heat period. Also, in British and American studies it was found that on average there is about a 90% level of egg fertilization from a single service, but generally only about 60 to 70% of these fertilizations result in a pregnancy. For mammals in general, about 20 to 40% of the fertilized eggs die during the first two to three days after fertilization, and in one instance when cows were slaughtered 23 days after insemination 25% of the embryos had died. It is thought that much of this failure to conceive is due to unfavourable gene combinations. Also, it has been claimed that the bovine embryo starts to be implanted on the wall of the womb about 21 days after conception and that it becomes fully established between days 35 to 42. Until then it is subject to stress-induced mortality and reabsorption. As cows become older the embryos implant less readily, probably because of small changes in the structure of the wall of the uterus. In one instance, the conception rate to the first service rose from 60.5% with heifers to 66.8% with 3-year-old cows and had declined to 51.1% after the eighth calving.

Several sets of results have shown that very fat cows and heifers can sometimes be slow to conceive. In one study, when cows were kept fat throughout the year or allowed to lose weight for part of the year and then recover what had been lost, the proportion still in the herd after 10 or 11 calf crops was 53% and 90% respectively. In another study, the average number of services required to become pregnant was 1.70 for fat cows and 1.43 for leaner cows.

Some results have indicated that the proportion of pregnancies from a single service can increase progressively with a longer time from calving in the first three months. One estimate was that the increase was likely to be from about 33% to about 69%, and in New Zealand records on 69,000 dairy cows artificially inseminated the increase was from 31% to 62%. Also, two sets of results indicated that the earlier cows come on heat after calving the more likely they are to conceive at the first service during the eventual mating period.

One claim has been that with artificial insemination the most effective time to inseminate is generally near the end of the heat period or about 12 hours after heat first becomes apparent.

PRODUCTIVE LIFE

Some reports show that individual beef cows can still be highly productive at 13 and 14 years of age. However, American results have indicated that, on average cows may remain in the herd for only about four to nine years after their first calving. These results clearly varied with the intensity of culling for failure to calve. The results are also likely to vary according to how the pasture conditions and the level of milk production affect the fertility level, together with the incidence encountered for dystocia and health problems such as bloat and grass tetany. One set of American results also indicated that overfeeding can result in a shorter productive life, and the earlier it begins the greater the effect is likely to be.

The effect of milking capacity on productive life was addressed in the chapter on milk production.

GESTATION PERIOD

The gestation period is generally longer with larger-type cattle and with tropical breeds compared to temperate breeds. When Angus and Hereford cows at Hamilton were mated to Hereford, Charolais, and Brahman bulls, the average length of the gestation period was 284, 288, and 291 days for the respective bull breeds. Other results for the length of the gestation period have been as follows: a difference of eight days between individuals of the same breed, an average of 282 days for heifers and 288 days for cows, one to three days longer for bull calves than for heifer calves, on average a reduction from 284 to 279 days with poor feeding, and 4 to 7 days shorter with twin calves than with singles. However, there can be some variations in these patterns, and at Werribee in Victoria the average gestation period was 2.5 days longer with heifer calves than with bull calves despite the heifers being lighter than the bulls by an average of 2.3 kg.

Bull Effect

To work well, the bulls must be free of deformity and other impairment and produce adequate volumes of viable semen.

Impairments can include arthritic joints, overgrown feet, and abnormally straight back legs that can restrict the bull's forward surge during mounting. Other problems can include a penis that is corkscrew shaped, is attached to the scrotum and so can't be extended, and is injured or infected.

Sperm production can be assessed according to the volume and density of sperm produced during an ejaculation and by the shape and vigour of the sperm. However, the results can vary appreciably between successive measurements on the same bulls. Tests on bulls under about 15 months of age are unreliable.

Bulls can differ substantially in their libido (willingness to serve), and in one instance when bulls differing in their assessed libido were mated to 50 heifers each for 30 days the resulting pregnancy levels ranged from 70 to 95%.

One method of testing for libido is to record how many times in 20 minutes a bull will serve a restrained heifer that is not on heat. It has been claimed that if a bull serves at least 10 times in this period it should be capable of serving about 75 cows, compared to only about 35 to 40 for a less-active bull. In Australia, this test was used by some stud breeders but now seems to have lost favour. American results have shown that while the test may have merit in identifying the bulls differing greatly in libido, the individual readings are only about 60% repeatable. Also, in another American study in which bulls were run in groups on the range and blood tests were used to identify the parentage of the calves, it was found that the most effective bulls appeared to be those with a medium score in the libido test. It was suggested that the bulls with greatest libido probably served individual cows so often that their semen levels became depleted.

The production of viable sperm requires an appropriate temperature in the testicles. To help maintain this temperature the testicles can move up or down in the scrotum. Testicles located within the body cavity generally don't produce viable semen.

A greater scrotal circumference measured at the thickest part of the scrotum has been associated with greater sperm production and generally better-quality sperm. It has also resulted in better overall breeding performance in the progeny, an effect that is moderately to highly heritable, with a reported heritability value of about 0.4.

In New South Wales, measurements of the scrotal circumference in 2-year-old beef bulls ranged from 26 to 43 cm. It was concluded that bulls with a scrotal circumference less than 30 cm are likely to be associated with relatively poor fertility, and that a scrotal circumference of at least 34 cm in these bulls is desirable. It was also claimed that bulls with a scrotal circumference of about 32 cm can achieve good fertility when joined to about 60 cows, and that bulls with a scrotal circumference of 35 cm can achieve good fertility when joined to about 75 cows.

A recommendation by the New South Wales Department of Primary Industries is that for British breed bulls the minimum desirable scrotal circumference at 12–14, 16–18, and over 20 months of age is 26, 29, and 31 cm respectively.

Canadian recommendations for a minimum scrotal circumference in fully grown bulls are the following:

Age (Months)	Simmental Charolais Maine Anjou	Angus Shorthorn	Hereford Blonde de Aquitaine	Limousin
12–14	33	32	31	30
15–20	35	34	33	32
21–30	36	35	34	33
Over 30	37	36	35	34

If adequately grown, young bulls can be used from about 14 months of age. One claim has been that sperm regeneration in these bulls is relatively slow, so it can be desirable to spell them after about three weeks. Also, if a bull is too small for the cows the bull can expel parts of its intestine while trying to serve the cows, and will require surgery.

Another claim has been that sperm takes about 60 days to fully develop, so bulls should be well fed for at least 60 days before mating. However, one assessment was that the effect of poor nutrition on the performance of bulls is generally from poor libido rather than low sperm production. Very fat and very lean bulls have been recorded as performing relatively poorly.

A bull will generally distribute its services across all the cows on heat at the same time, and if there is only a small number of cows on heat at one time the bull may serve individual cows two or more times.

In two sets of results, the conception rates were better when two bulls were used together rather than separately. However, in a large American study involving 19 herds in which bulls were used singly with an average of 28.7 cows per bull and 77 herds in which bulls were used together with an average of 18.7 cows per bull, the respective average calving rates were 87 and 81.3%. With multiple bulls, competition and fighting seemed to affect the result. One recommendation has been that if bulls are to be used together there is likely to be less fighting if the bulls are similar in age.

If bulls are used singly it can be desirable to rotate them between groups every three weeks and to inspect frequently for returns to service.

Various recommendations have been provided for the ratio of bulls to cows, and one recommendation is described above. The desirable ratio is likely to be affected by the age and vigour of the bulls and the adequacy of the pasture conditions to meet the nutritional needs of the bulls. In one study, when the number of cows per bull was increased from 20–30 to over 40, the average calving period lengthened from 77 to 132 days. Also, in a survey of 100 American herds of different sizes, the average conception rate was 96, 93.7, 94.1, and 88.9% when the number of cows per bull was 8–15, 16–30, 31–45, and 45–60 respectively, indicating a tendency for a moderate decline in the pregnancy rate when there was more than about 45 cows per bull.

A Canadian recommendation for the bull-to-cow ratio with a 60-day mating period has been as follows:

a) A bull should be at least 15 months old before being used and then should serve only about 15 to 20 cows, with the cows being brought to the bull.
b) Bulls should not run with the herd until at least 2 years old and then should be given only 20 to 25 cows to serve.
c) Bulls 3 years or older can be used on about 25 to 35 cows.

In the same Canadian recommendation, it was also claimed that, if well managed, cows should generally remain highly productive until 10 to 12 years of age and sometimes older.

In three studies, the pregnancy level increased with an increase in the length of the mating period in a range up to ten weeks. On the other hand, in a farm survey in Victoria and one in America there appeared to be no benefit from leaving the bulls with the cows for an especially long period to serve the cows that had failed to conceive when expected. One researcher claimed that with fit cows and bulls a nine-week mating period should be adequate.

Some farmers have calved twice a year to reduce the number of bulls required, but the effects on the dystocia levels and weaning weights need to be assessed (subjects addressed in other chapters).

Methods are available for inducing calving in cows more than 255 days pregnant, and for synchronizing when the cows come on heat. However, these practices are mainly used in dairy herds where artificial insemination is practised. An American set of results indicated that if the onset of heat is synchronized and the service is to be by natural means, it is desirable to have at least 4 bulls per 100 cows.

Miscellaneous Effects

Factors which have been implicated in reduced fertility in some instances include mineral problems, infectious diseases such as vibriosis and leptospirosis, and a high level of soluble protein in some clover-dominant pastures. In the past, poor rebreeding performance has sometimes been recorded in beef cows grazing lucerne for substantial periods, because of the oestrogen content of the plants. Also, in northern and some coastal regions of Australia a deficiency of phosphorus in the pasture can affect the fertility level by reducing pasture consumption.

As shown in the chapter on milk production, cows of greater milking capacity can require better pasture conditions if their fertility level is not to be reduced.

Hybrid Vigour

It has been found that the overall fertility rate can be especially responsive to the effects of hybrid vigour, with the degree of improvement generally being about 5 to 25%. One effect of hybrid vigour is greater robustness, so its benefits are usually greater under more-stressful circumstances.

Using different breeds of bull in rotation between different generations can be a way of obtaining hybrid vigour. In one study, the average fertility level was 65% with pure breeds, 75% with a two-breed rotation, 81%

with a three-breed rotation, and no further benefit from using more breeds. As to be expected, this indicates that for maximum benefit the cows should be cross-bred and the bulls of a breed not represented in the cows.

Sometimes crossbred bulls have been used to obtain some degree of hybrid vigour.

Multiple Births

Multiple births in cattle may arise naturally or can be induced by implanting fertilized eggs, or by injecting hormones to increase the number of eggs shed. With hormone injections the number of eggs shed can't be controlled.

Twins occurring naturally may be fraternal twins arising from the simultaneous fertilization of two eggs or identical twins arising from the splitting of a fertilized egg. On rare occasions triplets may be born and even more rarely quadruplets.

The frequency of multiple births can vary greatly between breeds, and it is generally higher in dairy breeds than in beef breeds. An American estimate of the likely average frequency of multiple births was 1.3% for Jerseys, 3.6% for Holsteins, 8.9% for Brown Swiss, 0.4% for Herefords, and 1.1% for Angus. It was also estimated that triplets are likely to born with a frequency of about 1 in 105,000 births and quadruplets with a frequency of about 1 in 665,000 births.

In a survey of 13 beef herds in Victoria in 1969 and 1970, the twinning rate varied from 0 to 4.28% between herds of predominantly Angus, Hereford, and Poll Hereford cattle.

A heifer calf born in association with a bull calf is known as a freemartin. On average, only about 5% of these heifers are fertile, because of fluid and tissue from the bull foetus in the womb affecting the development of the reproductive organs in the heifer foetus.

In an American study in which twinning was induced by embryo implant, it was concluded that the veterinary costs had increased by 40% with twinning and the profit level by 24%, but despite this increase in profit the economic merit of twinning was questioned. The dystocia level was 24.4% with single calves and 42.2% with twins, with most of the dystocia being due to calf size with singles and calf mal-presentation with twins. The total weight of calves weaned per cow at 200 days of age was 65.2% greater with twins. Only 4.3% of the female calves born as a twin to a bull calf were fertile.

In twinning experiments in Hamilton and Grafton, there was little effect of twinning on the dystocia level, but 8 to 17 % of the twin calves had to be assisted to stand or start suckling. Also, there were 12 to 25% retained placentas with twins, but only some of these cases required treatment.

In New Zealand, when a bull born as a triplet sired 44 daughters, 13 of them produced a total of 15 sets of twins and 6 of them each produced a set of triplets over their lifetimes. In an American attempt to increase the twinning rate in cattle, the average number of calves born per cow was increased from 1.07 to 1.29 over 13 years.

Conclusions

If heifers are adequately grown they can calve at 2 years of age, but the economics of the practice need to be considered. The mating of these heifers should begin before the main herd, and careful feeding is required to achieve favourable levels of fertility, dystocia, and rebreeding. Only those conceiving early should be retained.

Cows and heifers should be in good to moderately good body condition at calving, but over-fatness should be avoided. The milking capacity can affect the time taken to return to heat after calving.

Bulls differ in their serving capacity. It seems that a ratio of about 1 bull to 35–45 cows and a nine-week mating period are often likely to be satisfactory. Better results can sometimes be obtained by using bulls together rather than separately, but the bulls can sometimes fight. Bulls should be in strong body condition and free from impairment.

Diseases and mineral problems can sometimes also affect the fertility level. Commonly only about 60 to 70% of cows may conceive to a single service.

CHAPTER 13

Dystocia

The level of dystocia (calving difficult) encountered can appreciably affect herd productivity and profitability.

CALVING BEHAVIOUR

In managing dystocia it is important to know how cows behave at calving.

Almost all cows lie down during calving. In a Scottish study with 30 cows, 20 of them lay down until the calf had almost emerged, and then they stood up to complete the final expulsion. Nine remained lying down until calving was complete, and one stood up all the time.

In a Queensland study with Hereford cows which calved easily, the cows showed different degrees of restlessness, generally away from the herd, for a few days to a few hours before calving, and the udder and teats started to distend from a few days to a few hours before calving. Also, the cows arched their backs and slightly raised their tails from about three hours to about one hour before the water bag appeared. The average time (and the variation on each side of the average) from the bursting of the water bag until the calf was born was 66 (±16) minutes. Various times from birth were as follows: half an hour on average to the first calf movement,

69 (±18) minutes for the calf attempting to stand, 115 (±22) minutes for the first suckling, and 12.3 hours on average for the expulsion of the afterbirth with a variation of ±7.2 hours for male calves and ±6.4 hours for the female. Overall, there was a substantial tendency for a longer calving with heavier calves. Most of the cows licked the foetal fluid from the ground and ate the expelled placental membranes and the yellow faeces expelled by the calf.

Other results have shown substantial variation from these last results. In one instance, the average values for the duration of delivery were 116 minutes for heifers and 66 minutes for cows. In another report, the results for heifers v cows were 2–4 v 1–3 hours for the time of labour, and 1–2 v 0.5–1 hour for delivery. In this last instance, the contractions occurred about every 15 minutes to begin with and became progressively more frequent. The appearance of the water bag indicated the start of delivery.

An observation in several studies was that until the water bag appeared there was no clear indicator that calving was imminent. The use of external signs such as relaxation of the pelvic ligaments, the degree of udder tension, and cow behaviour were not reliable indicators of the likely onset of calving.

A general recommendation is that heifers should be left for three hours and cows for two hours before assistance is provided, unless there is evidence of some malfunction - such as only one foot exposed. With earlier intervention the pelvic ligaments may not have relaxed enough for the easiest of calving.

Most calves can remain alive up to four hours during the birth process and some may survive for as long as six to nine hours, but if the umbilical cord becomes pinched on the cow's pelvis and the calf's nose is not exposed the calf can die from a lack of oxygen.

Some cows become extremely aggressive as soon as the calf is born and remain so for about a day or so. Occasionally a cow will reject her calf and sometimes won't ever accept it.

The time taken to expel the afterbirth has differed between experiments, and generally most afterbirths have been expelled by 8 to 12 hours after calving. The entrance to the womb is usually tightly closed after about 36 to 48 hours, but closure can be delayed if there is retained afterbirth. The birth membranes are attached by cotyledons which are rich in blood vessels, and trying to remove them can result in critical bleeding if the removal is done incorrectly. The cotyledons can be eased off more readily if left for about 36 hours after calving. One recommendation is to remove any of the afterbirth protruding and insert antiseptic pessaries into the womb. Fluid dripping from the vulva indicates a prolonged problem, and a recommendation is to administer antibiotics. Retained afterbirths can be most common in the cases of abortion and twin births, but in a large American study there was no difference in the number of retained afterbirths between cows that had or had not experienced dystocia. There can also be increased afterbirth retention with selenium deficiency.

Calf Presentation

The onset of calving is initiated by the calf and is thought to be triggered by the degree of development of the calf's pituitary/adrenal gland system. Reported average gestation periods have commonly been about 278 to 286 days for temperate breeds of cattle and a few days longer for tropical breeds. The period can be relatively long with heavier breeds and with the use of bulls from these breeds, as well as usually being one to three days longer with the heavier birth weight that usually applies with bull calves compared to heifer calves. Also, it can be a few days shorter for 2-year-old heifers than for older heifers and cows, especially if the 2-year-old heifers are poorly fed. In addition, there can also be variations in the gestation period for reasons that are not clear. In Victoria, the average length of the gestation period with Hereford bulls mated to Hereford heifers was 281 days with one bull and 290 days with another, despite no significant difference in the average birth weight of the calves from these two bulls.

A calf is usually presented in a forward position with its front legs extended forwards and its head lying between them, or in a backwards position with its back legs fully extended. If the head or one or more of the legs is turned backwards this has usually to be corrected before calving can proceed. In extreme cases a vet will be required to remove the calf.

If a calf has to be pulled it should be pulled in a downwards arc. Working the calf backwards and forwards with calving ropes or chains attached to its legs can help to free the calf, and one vet claimed that a second person punching forcefully on the sides of the calf during difficult extractions can be beneficial. Inserting soapy water or calving fluid into the birth canal can help to lubricate the passage of the calf. Also, an antiseptic solution should be used to prevent infection if an arm has to be inserted into the womb to assist with the birth. Mechanical calf pullers are available, but if used too forcefully this can cause irreparable damage to the cow's spine. One experienced vet claimed that turning a cow on its back during a difficult calving can sometimes be highly effective in easing the calf's passage.

CAUSES

In numerous studies, about half the cases of dystocia have been associated with the size of the calf relative to the size of the pelvic opening in the dam, and the rest have been due to several other causes, such as abnormal presentation or deformity of the calf, the presence of twins, failure of the cervix and the vulva to dilate adequately, weak pushing by the cow or uncoordinated contractions, and injury to the cervix and birth canal from a previous calving. A deficiency of selenium or calcium has also been associated with increased levels of dystocia, and for cows kept indoors for long periods other associated factors have sometimes been a deficiency of vitamin E and a combination of gross fatness and lack of exercise.

A backwards presentation generally occurs in about 5–10% of births, and it can be associated with high levels of dystocia and calf death. Problems with this kind of presentation are unfavourable angles between the leg bones and the birth canal, the early presentation of the hips, the compression of the intestine expanding the ribcage, an increased likelihood of the umbilical cord being pinched against the pelvis of the mother, and possibly the lie of the hair.

The dystocia level can also be relatively high with twins. Causes can include the calves becoming jammed in the birth canal, the mother failing to strain after the first calf is born, or simply the delay in the second calf being born.

Other effects with twins are commonly a shorter gestation period and more retained afterbirths. In one experiment, when the first calving was at 3 years of age the results for singles v twins for calving difficulty were 20 v 54.4% at the first calving and 5.3 v 33.3% at the second. Also, the overall averages were 2.5 v 28.3% for retained afterbirths, and 281.8 v 274.8 days for the length of the gestation period.

In a report by vets on 635 cases of dystocia attended in Western Victoria, 67% of the cases were associated with a calf effect and 33% with a cow effect, and the number of cows that died was 37% of those with a calf effect and 13% of those with a cow effect. In this instance, the most common cow effect appeared to be failure of the cervix to dilate adequately, and in all these cases the cows were deemed to be too fat. In the 484 cases of a calf problem, 290 of them appeared to be due to oversized calves, 165 to head or limb displacement, and 29 to various deformities.

Age of Dam

Heifers usually experience more calving difficulty than cows, and the younger and less developed the heifer is at calving the greater is the likelihood of calving difficulty.

With Herefords in Queensland first calving at 2 or else 3 years of age, the values recorded for the respective age groups were 13.9 and 4.6% for the level of dystocia, 76.8 and 87.5% for the death rate in assisted calves, and 14.5 and 5.5% for the death rate near birth of the unassisted calves. The male calves were heavier than the females (30.7 v 29.3 kg), and the values for the respective sexes were 14.3 and 4.9% for the proportion requiring assistance and 21.2 and 12.2% for the proportion of the assisted calves that died.

In Britain, when the first calving was at 2 years of age the proportion of dams experiencing calving difficulty was 22.9, 3.4, and 0.8% at 2, 3, and over 3 years of age respectively. Different studies have shown a similar pattern with considerable variations in the actual figures. During the first few times of calving, the birth weight of the calf increases progressively with successive pregnancies, but this is countered by greater size in the cow, including a larger pelvic opening. In one instance, the average calf birth weight was 31, 33, and 35 kg for dams aged 2, 3, and 4 years respectively.

Bull Effects

When two breeds of cattle differing substantially in mature size are crossed, the birth weight of the calf differs according to which breed of dam gives birth to the calf. In one instance when there was a two-way cross between Angus and Hereford, the average birth weight was 30 kg with Angus mothers and 33 kg with Hereford

The use of larger and more muscular bulls can result in more dystocia, especially with heifers. In America, when Angus and Hereford heifers were mated to Jersey, Angus, Hereford, South Devon, Limousin, Simmental, and Charolais bulls, the level of dystocia generally increased progressively with larger-type bulls, and ranged from 13% to 74% with 2-year-old heifers, 3 to 28% with 3-year-old cows, and 0 to 11% with the older cows.

When cows in Britain were mated to Angus, Hereford, Limousin, Simmental, and Charolais bulls, the results varied progressively with an increase in the mature size of the bulls, and were 2.4 to 9% for assisted births, 1.3 to 4.8% for calf mortality, and 370 to 374 for the interval between successive calves. However, despite these adverse effects from using larger bulls the overall weight of calf weaned per cow mated increased from 179 to 208 kg with the use of these bulls. Clearly, the relative merit of using larger-type bulls in different situations will depend on how well the other management practices affect the level of dystocia.

HEIFER REARING

Several sets of results taken together have clearly indicated that with heifers to calve at 2 years of age it is usually best to have them well grown and in medium body condition at calving. This can benefit calving ease and rebreeding performance.

About 90% of the growth of the foetus occurs during the last 120 days of pregnancy and the feeding level at this time can sometimes influence the birth weight. Heifers and cows can use their body reserves to at least partially counter the effect of poor nutrition on the growth of the foetus. Generally the effect of the feeding level during the last third of pregnancy has more effect on the birth weight with leaner heifers, and there may be little or no effect on the birth weight with fatter heifers.

Two sets of results indicated that if 2-year-old, British-breed heifers are not gaining live weight at about 0.5 kg a day during the last 120 days of pregnancy they are likely to be drawing on their body reserves to support the developing foetus and surrounding membranes.

One of the most damaging practices can be to let the heifers lose weight during the last 120 days of pregnancy. This has resulted in smaller pelvic openings, weaker heifers and calves, more dystocia, more heifer and calf deaths, poorer calf growth, and poorer rebreeding. At Hamilton, when heifers to calve at 2 years of age were fed during the last twelve weeks of pregnancy at levels that resulted in average weight changes of -47, -11, and $+58$ kg, the results generally varied progressively with the level of weight change. The average results ranged from 26.7 to 29.9 kg for the birth weight, 18.4 to 51.6 cm^2 for the increase in pelvic area, 221 to 23 minutes for the heifer to stand, 306 to 86 minutes for the calf to stand, 7 to 1 for the number of calf deaths, 86 to 112 kg for the weaning weight, and 108 to 48 days to the first heat.

On the other hand, if heifers are very fat at calving then fat deposits in the birth canal can result in more dystocia, and in an American experiment a factor in the high dystocia rate with fat heifers was that most of the calves were presented backwards. Moreover, in a British result very fat heifers sometimes experienced health problems such as ketosis, and fat cattle can require more services on average to become pregnant.

Results obtained on a farm at Rutherglen with purchased, very lean but well-grown, drought-affected 3-year-old Hereford heifers indicated that with such lean heifers it may require little improvement in their nutrition in late pregnancy to greatly increase calf birth weight and the level of dystocia. When the heifers grazed on dry pasture, the early-born calves were unusually light and easily born. However, when the heifers were given access to a green-pick in a cereal stubble for 2½ weeks in the middle of the protracted calving period, the calves born afterwards were much heavier and the level of dystocia was horrendous. This kind of effect may be very important after a drought breaks

Bull and Other Effects

Bull calves are usually carried about one to three days longer than heifer calves, resulting in heavier birth weights and a tendency for more calving difficulty. Also, the hybrid vigour from cross-breeding can result in heavier calves at birth, but this has generally not affected the level of dystocia. In one instance, the increase in birth weight from hybrid vigour was 3.5% on average, and in another set of results the average birth weight was 1.55 kg greater for the crosses between Angus and Herefords than the average for the pure breeds.

With dairy heifers in Britain the use of bulls known to result in relatively light calf birth weights resulted in a reduction in the level of dystocia from 15 to 8%.

Because of the closeness of the relationship between birth weight, growth rate and eventual mature size, and the degree of imprecision in farm measurements of performance, some geneticists have questioned the merit of trying to select beef cattle for a lower birth weight without affecting their later growth rate. In a Canadian study, for each kilogram increase in birth weight there was a 2.86 to 4.42 kg increase in mature weight. In other studies, the relationship between birth weight and growth rate has generally been about 60 to 70% accurate, with differences in milk production and other factors which are difficult to identify probably affecting the results. However, in an American experiment conducted over 16 years, simultaneous selection for below-average birth weight and heavier yearling weight resulted in progress for both attributes, but the progress in selecting for heavier yearling weight was only 61% of that achieved by selecting for yearling weight alone.

In one instance, when Angus bulls of EBV (estimated breeding value) −0.95, −0.82, +2.9, and 2.7 kg for birth weight were mated to crossbred-heifers, the average birth weight was 33 kg for the progeny of the two bulls of lower EBV and 36.1 kg for the progeny of the others. However, the range in birth weight was as wide as 27 to 45 kg for the progeny of the individual bulls.

In another experiment, when Angus and Hereford heifers to calve at 2 years of age were each mated to four Angus and four Hereford bulls, the differences in the levels of dystocia from using the individual bulls seemed to be due mainly to the difference between the bulls in the range of birth weights in their calves and not to any difference in the average birth weight of the calves sired by the individual bulls. The range in the average level of dystocia was 20 to 44% with the Angus bulls and 13 to 40% with the Hereford bulls.

In one study, a difference between the calves in their head and shoulder widths seemed to affect the level of dystocia, but a similar effect has not been found in some other studies when allowance was made for the differences in the birth weight.

In commercial herds, since the main dystocia problem can be with 2-year-old heifers, a better practice than trying to select for a lighter birth weight may simply be to use smaller-type bulls from breeds such as the Galloway and Belted Galloway for serving heifers. This also reduces the problem of heavy bulls serving light heifers. The use of bulls that will colour-mark the progeny will also be a means of identifying these progeny.

In American results in which embryo implants were used to achieve cows bearing single or twin calves, the level of dystocia was 20.4% for the cows bearing single calves and 42.2% for those bearing twins, and the survival level to 200 days of age was 15.2% higher with the singles. Only traction was required in 85% of the cases of dystocia with single calves whereas the presentation had to be corrected in 78% of the cases with twin calves, and of these 59% also required traction.

In pure-bred commercial herds, the attempts to achieve faster growth rate by using faster-growing bulls has appeared to result in more dystocia in some instances but not in others. The variation may have been due to the degree of difference in the growth capacity between the successive bulls and differences in the feeding practices between the herds affecting the calf birth weight.

Miscellaneous Effects

In Britain, a lack of exercise from having cows tethered for long periods during winter has been associated with some increase in the dystocia level However, it seems that grazing cattle usually experience adequate exercise for calving ease. In an American study, making heifers walk several kilometres each day to and from hay-feeding rather than just providing the hay in the home paddock had no apparent effect on the ease of calving.

In Victoria, results indicated that greater disturbance at calving may increase the likelihood of dystocia. In this experiment, the average level of dystocia over the first three calving periods was 10.3% for those calved in a paddock compared to 36.8% for those experiencing greater disturbance from calving in a yard. However, with the heifers in a yard, the closer inspection may have prompted earlier intervention than was desirable.

In one set of British results, very fat cows seemed to experience more dystocia. When the cows were in condition score 2, 2.5, 3, 3.5, and 4 on a scale of 0 to 5 (lean to fat), the corresponding range in the dystocia level was 5.3, 6.8, 7.1, 8.5, and 12.2%, and in New Zealand results having heifers fat when calving at 2 years of age resulted in more dystocia and lower milk production. On the other hand, in another set of British result for cows mated to Charolais bulls and calving in condition score 2.5–3 or 3.5–4 on a scale of 0 to 5, the level of dystocia was 10.4 and 7% for the respective groups. Also, in one British experiment any tendency for fatter cows to experience more dystocia was partly countered by these cows giving birth to lighter calves.

In some studies conducted to examine the relative importance of different factors on the incidence of dystocia in cows, the body condition has not been a significant influence, but this result may have been affected by the range in body conditions that applied.

In experiments on stocking rates and calving seasons of Angus at Rutherglen, the use of different stocking rates resulted in a large variation in the body condition at calving but had no obvious effect on calving ease. In this instance, the apparent effects on the dystocia level were heifers compared to older cows and the presence of green pasture at calving affecting the calf birth weight.

When cows in America were kept fat throughout the year or allowed to lose weight for part of the year and then regain it, as usually applies on a farm, there was a tendency for more dystocia with the persistently fat cows.

Overall, therefore, the effects of fatness on the incidence of dystocia have sometimes been a bit variable. However, apart for a likely effect on dystocia in some situations, it is a waste of feed resources to allow breeding animals to become grossly fat. Changes in the stocking rate, milking capacity, level of supplementary feeding, and the calving season can be means of varying the body conditions encountered.

Pelvic Opening

Considerable study has been made on the value of measuring the size of the pelvic opening for predicting the likelihood of dystocia in 2-year-old heifers. Measurements taken just before mating and just before calving have shown a close relationship between the two sets of measurements, indicating that it can be appropriate to take these measurements at either of these times.

The overall conclusion has generally been that these measurements can be of value for identifying any heifer with an abnormally small pelvic opening but not for identifying the likelihood of calving difficulty for individual heifers. This situation occurs because the birth weight of calves from heifers is generally proportional to the size of the heifer, so a small heifer with a small pelvic opening can give birth to a small calf that results in no dystocia.

In one instance, the use of measurements of the size of the pelvic opening could predict about 12% of the cases of dystocia when allowances were made for the differences in the heifer live weight.

Conclusions

Dystocia can be caused by various factors of which the most important is usually the birth weight of the calf relative to the size of the dam.

The problem is much more common with heifers than with cows, especially with younger and less-developed heifers. Increased levels of dystocia can arise from having the heifers fat at calving, gaining or losing too much weight during the last third of pregnancy, and being mated to larger-type, more-muscular bulls.

With cows, the level of dystocia can be affected by the use of larger-type and more-muscular bulls, the level of feeding during the last third of pregnancy, and persistent fatness. It can also be especially high with the birth of twins and with calves presented backwards.

CHAPTER 14

Breeding Principles

In animal breeding it is important to know the breeding principles that apply and how they are likely to affect the results obtained.

GENES

How an animal develops and functions is under the control of its genes, which are located in chromosomes in a dense area of each cell called the nucleus. These chromosomes are ladder-like structures in which the sides are composed of phosphates and sugars and each rung is composed of a combination of the compounds adenine and thiamine or cysteine and guanine. The sequences of the rungs of different chemical compositions located on the chromosomes form the animal's genetic code, and chemical messages sent from the genes control the form and functioning of the body. The overall chemical structure of the genes is called deoxyribonucleic acid, or DNA for short.

The messages which can be sent by this system of two different chemical combinations are similar in complexity to what could be sent by the old Morse code method and by the programmed sequence of on–off switches in the modern computer.

There are 30 pairs of chromosomes in cattle, 2 of which are known as the sex chromosomes. Females have two pairs of what is called the X chromosome, and males have a pair each of the X and Y chromosomes. Thus, an attribute contained on an X chromosome can be transmitted to either a son or a daughter, but one contained on a Y chromosome can be transmitted only to a son.

In each cell nucleus, the chromosomes occur in pairs in which the individual chromosomes are wound round each other. During sperm formation in the testicles or egg formation in the ovaries, the two chromosomes in each pair separate and move at random to opposite sides of the cell from each other. A wall then forms down the middle of the cell to produce two sperms or two eggs, each with half of a normal complement of chromosomes. At fertilization, a sperm and an egg fuse and a full complement of chromosomes is restored.

It is because of this random movement of the chromosomes during sperm or egg formation that no two progeny from the same parents are usually identical, unless they are identical twins resulting from the division of a fertilized egg.

On extremely rare occasions there may be a spontaneous change in a part of the gene structure of an animal that is not directly due to the mating practices. This is called a mutation, and it can change the animal's characteristics to varying degrees. Also, some part of a chromosome can break off and combine with part of another chromosome, and thereby have some effect on the gene structure. Moreover, the genetic material contained in eggs or sperm can become damaged as an animal becomes older, but this factor does not seem to be greatly important in cattle.

Aside from the results of normal breeding, there are techniques for inserting outside genetic material into the nucleus. This method is used extensively in plant breeding but has been little used in cattle breeding. In addition, an animal can be cloned to produce offspring identical to itself, but this practice has been fraught with health problems in the progeny.

Animal Attributes

The degree to which an animal attribute can be changed by selection depends on the degree of variation there is to select from and how many factors determine the result. For example, the growth rate could be influenced by such factors as the amount of milk consumed, the potential mature size, the resistance to parasites, the level of appetite, and the tolerance to the climatic conditions. The approximate degree of success likely is assessed in breeding experiments, and the results are scored on a basis of 0 to 1, with 0 indicating that any success is unlikely and 1 indicating that complete success is likely. The heritabilities, as these values are called, can be negligible for attributes where natural selection operates continually, such as in fertility because animals of low fertility reproduce poorly or not at all. However, while direct selection for improved fertility can be unrewarding, some factors associated with it can be selected to some degree, such as birth weight.

To demonstrate the use of heritabilities, suppose that the average growth rate in herd A is 10% greater than that in herd B and that the heritability for growth rate is 0.4. If a representative sample of bulls from herd A is to be mated to a representative sample of cows from herd B, then the average growth rate of the progeny can be expected to be about $10 \times 0.5 \times 0.4 = 2\%$ greater than that of herd B. This occurs because the bulls contribute only half (0.5) of the genes of the progeny, and these will be passed on with an effectiveness of only about 40% (0.4). It must also be realized that there will be variation in the production merit between the different bulls in herd A and that a close examination of the performance of their progeny and relatives will help to refine the assessment of the individual bull's value.

Some of the reported heritabilities have been 0.15–0.20 for calving ease in heifers, 0.20–0.25 for milk yield, 0.25–0.50 for growth attributes, and 0.25–0.50 for temperament.

Selecting for fewer attributes simultaneously is likely to enable faster progress because it will involve fewer compromises.

GENE TYPES

Genes can be dominant, recessive, moderating, or linked to sex, and for some attributes the result can depend on the combined effect of several genes.

A dominant gene masks the influence of other genes. Examples of dominant genes are those for the black coat colour in Angus cattle and the white faces of Herefords.

All animals have harmful recessive genes, but their presence remains hidden until two of them come together in the absence of more dominant genes. Inbreeding increases the likelihood of this occurring, and one of the most damaging effects of inbreeding can be a substantial decline in the fertility level. Also, an example of the harmful effect of inbreeding was evident in the past when intensive selection for small, blocky cattle resulted in the appearance of dwarf cattle with big heads and short legs. More recently, harmful recessive gene effects have been evident in intensively selected American herds.

An example of the effect of moderating genes is in the coat colour of Shorthorn cattle. In these cattle there is a gene for red coats and one for white coats, and the combination of the two genes in one animal results in a roan coat. If roan-coated animals are mated to each other there is a rearrangement of the genes, resulting in the proportion of the progeny with a different coat colour being about a quarter red, a quarter white and half roan.

CROSS-BREEDING

Cross-breeding can be used to combine the production attributes of two breeds, and by the action of what is called hybrid vigour it can make animals more robust.

Greater robustness in the animal enables it to cope better with unfavourable situations. As a result, the actual benefit obtained from cross-breeding can vary according to the severity of the environmental challenges experienced by the animal. In a favourable environment a notable benefit from cross-breeding may only be evident in something like the fertility level following a year of poor pasture growth, whereas in a less favourable environment there can be more continuous benefit. Also, the degree of hybrid vigour is usually greater with greater genetic diversity between the breeds used, and it can be especially great in crosses between temperate and tropical breeds.

The degree of hybrid vigour is measured by comparing the performance of the progeny with the average for the pure-bred parents. The benefit is usually greatest for the overall fertility level, and the average values for several results reported were as follows:

Attribute	**Benefit from Hybrid Vigour (%)**
Calving Rate	10
Calf Survival	4
Weaning Weight	6
Total	20

For the maximum hybrid vigour, the cows need to be crossbred and the bulls need to be from of a breed not represented in the cows. American results indicated that if rotational cross-breeding is used there is likely to be little additional benefit from using more than three bull breeds. In this instance, the fertility results achieved from pure breeding or a rotation with two, three, or four bulls were 65, 75, 81, and 81% respectively.

The degree of inbreeding in a particular herd may not be readily apparent, and the closeness of the relationships between animals may be hidden in past pedigree records. In an investigation into the cause of the poor fertility in Dorset Horn sheep in Australia, a CSIRO researcher found that most of the breed was descended from a single ram that had been used extensively decades earlier.

There can be practical and financial problems in cross-breeding on individual farms, and these have to be considered carefully. Generally the practice is easier to adopt in smaller herds where all the replacement breeding animals are purchased. The use of crossbred bulls can be one way of achieving a measure of hybrid vigour.

Associated Effects

In trying to improve an animal attribute it is important to consider possible associated effects that may occur. For example, birth weight, growth rate, and mature size are closely related, and as described in another chapter, some geneticists questioned the possible overall merit of selecting for lighter birth weight. However, it was also reported that selecting simultaneously for lower birth weight and greater yearling weight in one experiment achieved progress in both attributes, albeit at a slower rate for each.

Similarly, in an American study, selecting for a larger eye muscle proved to be largely just selecting for larger-type animals, and in another experiment selecting for faster growth also resulted in larger-type animals. It may be, therefore, that in selecting for more-muscular cattle the eye muscle area has to be considered in association with growth rate or mature size.

An American researcher also claimed that selecting for ever-faster growing cattle to meet the preferences in feedlots had adversely affected some maternal factors, such as the gestation length and age at first puberty.

Overall, therefore, in intensive selection for one attribute the possible consequence on another needs to be considered and effective counteracting measures included.

Breeding Merit

With the beef cattle recording schemes operating in Australia, estimated breeding values (EBVs) are given for various animal attributes and presented as the degrees of difference from overall base values. They are obtained by using records from different relatives and weighting the results according to the closeness of the relatives. Their accuracy depends on how many records are available for the particular animal, and the estimated accuracy estimates are reported.

In obtaining the EBVs, the records are adjusted for various factors that are likely to have influenced the results. These factors include the season of birth, cow age, and the overall level of nutrition. To assess the degrees of adjustment required in individual herds, sample matings are made by AI using bulls of known genetic merit. Also, milk production is estimated from the growth of the calves to weaning compared to their later growth.

The likely outcomes of any mating cannot be predicted accurately because of some of the reasons already described, but EBVs provide the best predictions possible, and their overall value has been proved.

The animals should be selected for their EBVs according to whether the desire is to increase, stabilize, or reduce a particular attribute. For example, it may not be desirable to keep selecting for faster-growing animals if it is mainly going to lead to larger-type animals, and instead the desire may be just to stabilize the mature size of the animals one already has.

Where change is desired it can be achieved more quickly for the herd as a whole by more rapid turnover of the generations. However, where change is not required there can be merit in retaining the older animals of proven performance, because these animals are likely to have been the most profitable ones in the herd. Aside from that, a reduced turnover reduces the management costs and increases the number of progeny that can be available for sale, as was described earlier.

In addition to reporting the EBVs, some breeders now also include a structural assessment of cattle being sold, with the assessment having been done independently according to defined standards.

Genetic Testing

Genetic testing can now be done to confirm the parentage of individual animals and to detect for some harmful and beneficial gene combinations. This technology is likely to develop rapidly.

Conclusions

Animal breeding can be complex, and in trying to achieve progress it is important to understand the mechanisms involved. The use of EBVs can be an effective way of maintaining or changing particular production attributes, and cross-breeding can be highly beneficial in some circumstances.

The overall aim in any breeding programme should generally be to attain animals with attributes well suited to the particular production situation.

CHAPTER 15

Animal Selection

It is important to use animals with appropriate characteristics relative to the pasture conditions that will apply and the standard of carcass required. The pasture conditions affect the growth rate of the animals and the standard of carcass it is practical to produce at different ages. Some of the important animal attributes are described below.

ANIMAL SIZE

Larger-type breeding animals produce faster-growing progeny that are heavier at a given age but less mature and hence leaner at a given weight. If the breeding animals are too small or too large for producing carcasses of the preferred weight, then the carcasses may be too fat or too lean respectively at that weight. This could involve selling the animals earlier in the year before best use can be made of the pasture, or else keeping them to an older age or having to use uneconomic levels of supplementary feeding. Also, using larger-type breeding animals to achieve a certain progeny weight at a younger age can increase the cost of keeping the breeding herd relative to the weight of progeny sold.

In the results of a computer model constructed to examine the effects of mature size and breed on the thickness fat cover at the same carcass weight, the average thickness of fat cover predictions for a carcass of

180 kg chilled weight were 10 mm for Herefords with average mature weight of 650 kg, 7 mm for Herefords with average mature weight of 750 kg, and 5 mm for Charolais x Hereford. It was also predicted that the larger-type Herefords would grow about 5% faster than the smaller type.

For markets such as the home market, carcasses of moderate weight are generally preferred, so in producing these carcasses breeding animals of moderate size are likely to be best. On the other hand, for servicing some overseas markets heavy carcasses are required, so in producing these carcasses it is likely to be best to use larger-type cattle. Because of the number of outlets for Australian beef a range of carcass attributes can be accommodated, but the weight and fatness of the animal will affect the price it attracts, as is evident from market reports and the price schedules applying in directly selling to abattoirs. However, it is important to note that over time the carcass preferences can change, and the carcasses now preferred for the home market are heavier than was the case earlier.

One way that animals of greater mature size can be used to improve efficiency is to use bulls of a larger type than the cows. This increases the weight gain of the progeny relative to the feed cost of keeping the cows. However, in using this practice it becomes more important to use management practices likely to moderate the incidence of dystocia.

As a point of general interest, the desire to achieve larger-type cattle has been driven from America where an aim has been to reduce the production and slaughtering costs per unit of carcass. It has been estimated that over the 40 years to 2013 some average increases have been about 40% in cow size, 27% in overall feed requirement, 200% in peak milk production, and 16% in feed requirement to support the peak milk production. Much of the beef in America seems to be produced in a three-stage system: (1) breeding on rangelands, (2) rearing the weaners by stockers, and (3) finishing in feedlots on high-grain rations. For the stockers and feedlotters in this system the benefit from using faster growing cattle that gain more weight can be beneficial. Moreover, the grading system seems to favour the substantial fatness that can result from feedlotting.

Muscling and Fatness

The relative proportions of bone, muscle, and fat in the carcass affect the value of the carcass to the retailer or meat processor. If the carcass is not fat enough it can be unsuitable for sale in some markets, and if it is too fat there has to be wasteful trimming of the surplus fat. Also, if the carcass is too large it may result in some cuts being larger than desired

Aside from being affected by the degree of animal maturity, the composition of the carcass can be greatly affected by the breed or breeds used and sometimes by the method of feeding, as was described in the chapter on growth and development.

There are three general groups of temperate-breed cattle used to varying degrees in beef production which can appreciably affect the standard of carcass produced. These are the following:

a) *The early-maturing beef breeds.* These are cattle of small to moderate mature size and include most of the common British breeds, such as the Angus, Hereford, and Shorthorn. They are considered as early-maturing because they fatten at relatively light weights, partly because of a smaller mature size and partly because of past selection for fattening ease.

b) *The late-maturing breeds.* These are relatively large types of cattle and include the European breeds, such as the Charolais, Limousin, Simmental, and the British breed South Devon. They are regarded as late-maturing because they fatten at relatively heavy weights. Also, they generally deposit a greater proportion of their fat in the body cavity rather than under the skin.

c) *The dairy breeds.* The dairy breed most commonly used to some degree in beef production in Australia is the Friesian. The dairy breeds are relatively lightly muscled with a comparatively low proportion of muscle to bone, and on pasture they can be difficult to fatten until they are approaching maturity. Also, when sold they can be heavily penalized on price. They have similar fat distribution patterns to the late-maturing breeds but tend to deposit subcutaneous fat just a little more readily than these breeds. In beef production, the dairy breeds and large European breeds are mainly used as crossbreds.

The use of different breeds or breed combinations can be most effective in changing the standard of carcass produced.

The following American results from the 1970s show the relative merit of using bulls from different breeds as terminal sires. The bulls were used on Angus and Hereford cow, and the results for the progeny were corrected to the same time in a feedlot.

Bull Breed	Carcass Weight (kg)	Fat Cover (mm)	Eye Muscle Area (cm^2)	EMA/ Carcass Weight
Angus	288	14.2	76.1	.264
Hereford	288	13.2	75.6	.263
Friesian	292	8.6	74.3	.254
Brown Swiss	298	8.9	81.7	.274
Simmental	302	8.3	82	.272
Limousin	293	8.7	84	.287
Charolais	310	10	84	.271

These results show the lighter and fatter carcasses that applied from using the British breed bulls, and overall the generally thinner muscling relative to the carcass weight from using the Friesian bulls. The values for eye muscle area divided by carcass weight provide a general indication of the differences in the proportion of muscle in the carcass. The British breeds have a lower proportion of muscle in their carcasses because of a higher proportion of fat.

Of the most popular European breeds of cattle in Australia, the Charolais and Simmental are generally the largest animals, Limousin has a relatively good muscle-to-bone ratio, and milk production is good in Simmental cows and poor in Charolais and Limousin.

In another piece of American information, the following ratings were given in the past for the production attributes of some of the most popular breeds there.

Breed	Growth Rate / Mature Size*	Ratio of Lean to Fat*	Age at Puberty*	Milk Production*
Angus	++	++	+++	++
Hereford	++	++	+++	++
Red Poll	++	++	++	+++
South Devon	+++	+++	++	+++
Pinzgauer	+++	+++	++	+++
Brahman	++++	+++	+++++	+++
Gelbvieh	++++	++++	++	++++
Simmental	+++++	++++	+++	++++
Limousin	+++	+++++	++++	+
Charolais	+++++	+++++	++++	+

*More crosses indicate faster growth/greater mature size, leaner carcasses at a given carcass weight, older age at puberty, or greater milk production. These ratings were basically provided to aid in deciding which breeds to use in cross-breeding.

The above relationships are still likely to apply generally in Australia, but they have now been superseded in America to at least some degree. In preliminary results on breed assessment conducted about 25 years later in 1999 and 2000 at the Roman L. Hruska US Meat Animal Research Center in Nebraska University, it was found that any growth rate and carcass weight differences between British and European breeds had virtually disappeared, because with British breeds there had been intense selection for faster growth and greater mature size, whereas with the European breeds the selection emphasis had been on a lighter birth weight and easier calving.

In this last experiment, bulls from various breeds were mated to Angus and Hereford cows that were at least 4 years old. Pre-weaning results were reported for two years and post-weaning results for one year. These results were corrected to a 448-day slaughter age and for the absence of hybrid vigour when the Angus and Hereford bulls used were mated to cows of their own breed. The results were as follows:

Sire Breed Averages for Pre-Weaning Traits for 1999 and 2000[a]

Sire Breed of Calf	Gestation Length (d)	Unassisted Calvings (%)	Birth Weight (kg)	Survival to Weaning (%)	200-d Weaning Weight (kg)
Hereford	284	95.6	41.1	96.2	238
Angus	282	99.6	38.1	96.7	242.1
Red Angus	282	99.1	38.4	96.7	238.9
Simmental	285	97.7	41.9	96.7	251.1
Gelbvieh	284	97.8	40.3	97.1	242.5
Limousin	286	97.6	40.6	96.9	235.7
Charolais	283	92.8	42.6	97.1	245.2
LSD[b]	1.5	3.4	1.4	3.8	6.36

[a] Germplasm Evaluation Program Progress Report No. 21.
[b] Breed differences that exceed the LSD are significant ($P < 0.05$).

Sire[a] Breed Averages for Post-Weaning and Carcass Traits for 2000

Sire Breed of Calf	Post-Weaning ADG (kg/d)	Slaughter Weight (kg)	Carcass Weight (kg)	Marble Score (%)	USDA Choice (%)	Yield Grade	Fat Th. (mm)	Eye Muscle (cm²)
Hereford	1.57	618	377	538	79.1	3.35	14	82.2
Angus	1.54	624	384	577	93.6	3.32	14.7	87
Red Angus	1.54	618	381	589	96	3.76	15.2	78.8
Simmental	1.57	630	387	536	61.2	2.95	10.7	88.5
Gelbvieh	1.51	611	375	514	63	2.80	10.1	90.5
Limousin	1.50	593	370	507	44.8	2.63	10.4	90.5
Charolais	1.56	621	382	517	75.7	2.77	10.9	90.4
LSD[b]	0.08	25	15	35	22.5	.41	2.8	4.8

[a] Germplasm Evaluation Program Progress Report No. 21.
[b] Breed differences that exceed the LSD are significant ($P < 0.05$).

These results show that despite the increase in the size of the British breeds, they still fattened more readily subcutaneously than the European breeds, and the Angus and Red Angus still marbled relatively well. Collectively the British breed bulls produced progeny which graded 88.8% choice or higher and 22.3% were of yield grades 1 and 2. The continental-breed sires produced progeny with carcasses that were 60.9% choice or higher and 57% were of yield grades 1 and 2. The American grading system favours fat carcasses with marbling.

Overall it is a question of how large-type animals will fit into the particular production situation, including assessment of any likely effect on the prices received when the cattle are sold. In some overseas markets for Australian beef there is also a strong preference for carcasses with substantial marbling. Angus cattle are currently favoured for servicing some of these markets. The Japanese Wagyu breed has been strongly selected for ease of marbling.

How the animals are sold can affect the standard of carcass it is most profitable to produce. There is strong evidence that buyers of prime stock in saleyards tend to favour relatively fat cattle because of a higher dressing percentage and perceived thicker muscling. On the other hand, if the animals are sold on carcass measurements at an abattoir the assessment is likely to be more closely aligned to commercial value.

In Australia, the dependence on pasture for beef production and the variable pasture conditions between years, together with the generally poor return from supplementary feeding, affect how practical it is to try to be too precise in the standard of carcass produced. Supermarkets are now sometimes finishing some cattle in feedlots to obtain the product preferred.

Milk Production

As was described in the chapter on milk production, it is most important to use cows with a milking capacity suited to the pasture conditions that will apply.

Calving Ease

It can be highly beneficial to use breeds or breed crosses which result in relatively easy calving. This subject was addressed in the chapter on dystocia.

Miscellaneous Factors

Selecting for a quieter temperament and absence of horns can make the cattle easier and safer to work with and reduce the likelihood of bruising and dark cutting meat in the carcasses. In three Australian trails, the average amount of cattle trim because of bruising was 0.77 kg with polled cattle and 1.59 kg with horned. British breed cattle are now generally of satisfactory temperament, but even within some of these breeds, there are strains with poor temperament. There can be a substantial difference in temperament between the large European breeds.

Cattle can be dehorned but it is a painful and unnecessary task, and there is no evidence that even within the same breed polled cattle are inferior in any way to horned cattle.

In the case of white-faced cattle, the risk of eye cancer can be reduced by selecting animals with pigmented eyelids.

In Queensland, pink-eye infection resulted in a 10% average reduction in growth rate with the incidence of this disease being 43.1, 21.4, and 7.2% for Hereford, Simmental, and Africander x Hereford cattle respectively. This illustrates the generally high level of resistance of tropical breeds to this infection. Also, in America the incidence of pink-eye infection ranged from 0 to 63% in the progeny of the different sires, indicating a strong genetic influence on the occurrence of the disease, but selection for improvement may be difficult other than not buying a bull showing evidence of infection.

There can also be merit in selecting for anything that may impair the eventual functioning of the animals, and some breeders are now providing assessment of structural factors based on defined standards.

Conclusions

Overall, cattle should be selected on the basis of their suitability for producing carcasses of the desired standard under the feed conditions it is most economical to provide. Especially important attributes are the mature size, muscling and fattening characteristics, level of milk production, ease of calving, and temperament. Cross-breeding can be useful in achieving the desired results.

CHAPTER 16

Beef Quality

The appearance and eating properties of beef can affect its attractiveness to customers.

The eating properties of beef are usually assessed on the basis of tenderness, flavour, and juiciness by trained tasting panels given samples with a standard size and degree of cooking. The tenderness can also be assessed according to the weight that has to be applied in a Warner–Bratzler machine to cut through cores of meat for a standard size and degree of cooking. The two methods of measuring tenderness generally provide similar but not identical patterns of results.

Tenderness is usually the most important quality attribute, but the relative importance of the different attributes can differ between the markets in different countries, probably because of differences in the cooking methods.

Appearance

The surface of freshly cut meat is purple in colour, and with exposure to the air the colour changes to bright-cherry red. With prolonged exposure, the colour can become brownish. Also, if cattle are killed at a high body temperature this can result in a rapid fall in the pH level of the meat after the animal is slaughtered, resulting in meat that can exude moisture when on display, and some of the practices described below can result in dark-coloured meat. Moreover, cattle in a feedlot for a substantial period can suffer from a vitamin E deficiency, and this can affect the colour of the eventual meat and cause the fat to become rancid.

The muscle colour is usually paler and the fat is whiter in the meat from cattle intensively fed on grain rather than grazed. Also, the muscle colour of the meat from younger animals is usually lighter in colour than that from older animals, because of a lower concentration of the pigment myoglobin in the meat.

The texture of the cut meat surface can vary, and generally a fine, velvety texture is sometimes regarded as being likely to indicate a slightly more-tender meat than a coarser texture. In general, older cattle have larger muscle bundles, and as a result have coarser meat. In older cattle it seems that any reduced tenderness is due to the age effect and not to the texture effect, and between cattle differing in age there can be texture difference without any difference in tenderness.

The yellowness of the fat from grazing cattle is due primarily to the presence of the pigment beta-carotene, which is the base substance for the formation of vitamin A. Beta-carotene is present in fresh green plants, and the concentration usually declines as the plants mature. It is also present to varying levels in hay and silage, with the concentration usually being higher in silage than in hay.

The fat colour tends to become more intense in cattle as they age beyond about four years. There also appears to be a strong genetic influence on the fat colour, which probably acts through the level of absorption of beta-carotene from the gut, and can result in variation in the fat colour between individual animals of the same breed. Moreover, the fat in some dairy breeds, especially the Jersey, has tended to be yellower than that in some beef breeds.

Feeding cattle for a period on a diet low in beta-carotene can result in whiter fat, especially if the animals are intensively fed on grain for about 90 days or more. This whitening occurs partly because virtually all the tissue within an animal's body is continually being broken down and renewed rather than remaining as it was when first deposited.

The meat from bulls is usually darker in colour than that from steers and heifers, and if the bulls are not fairly young the meat can have a bull flavour. The meat from bulls is also more variable in tenderness.

Tenderness

Beef is more variable in tenderness than other popular meats.

The factors affecting meat quality are complex and not fully understood, but some influences have been identified.

The muscle component of the carcass is composed of long thin cells, called myofibrils, bound together by connective tissue into bundles and further bound to form individual muscles which are connected to bones by tendons. In the cells are interleaving fine filaments composed either of the protein myosin or actin. In the middle sections of the cells, filaments of different protein composition can move over each other under nerve stimulation, causing the muscle to relax or contract.

After slaughter, the muscle fibres contract and the muscles shorten. If the muscle shortening occurs at moderate temperatures the process is called rigor mortis, and the claimed completion times in beef carcasses have generally been from 20 to 48 hours. During rigor mortis, glycogen present in the muscle is converted into lactic acid and ideally the pH level of the muscle should fall over 24 hours from an initial value of about 6.8 to 7.2 to one of about 5.4 to 5.7. On the other hand, if faster cooling results in the meat temperature falling below about 12 °C before the muscle pH has dropped to below about 6 to 6.4 and if there is still sufficient reserve energy in the muscle, then cold shortening of the muscle can result in tough meat that is unresponsive to the effect of later ageing. Furthermore, if the ultimate pH is reached before the carcass has cooled sufficiently, heat shortening can result in pale and watery meat and reduced tenderness because of impairment of the tenderizing enzymes.

If cattle experience severe stress or prolonged starvation just before slaughter, their body reserves of glycogen (animal starch) can become too depleted for an adequate level of lactic acid to develop after slaughter. This can result in tough, dark cutting meat. In one report, the pH of this kind of meat was 6.8 and the shear value was four times than that of a meat with pH 5.7. Aside from having the cattle in unfamiliar surroundings, increased stress can occur from such factors as the mixing of groups, cows on heat, rough handling, presence of cattle with horns, and severe cold, wet weather. On long journeys and with prolonged fasting the blood glycogen levels can decline slowly. Generally, the loading density of trucks has little effect on the degree of stress unless the density is great enough to increase the risk of foundering.

In Danish results with young cattle, the proportion of dark cutting carcasses and a muscle pH of 6.2 or above was 2% in animals slaughtered when they arrived at the abattoir and 9% in those held overnight before slaughter or delivered from a saleyard.

For cattle that have been off feed for a substantial period it can take at least two days of feeding to restore the muscle glycogen level significantly, and it may take at least four days to restore it substantially. Often there can be little benefit from feeding cattle at abattoirs because of the inadequate nutrition and a continuing stressful situation. Cattle that have come off grain-feeding are less likely to be deficient in glycogen.

The degree of dark cutting in the meat can vary between different muscles in the carcass.

If the meat is frozen before rigor mortis is complete there can be a substantial shortening of the muscle fibres after thawing, and a substantial reduction in weight because of moisture weep.

Heavier and fatter carcasses and the thicker parts of the carcass cool more slowly, so they are less subject to cold shortening.

At abattoirs there is an incentive to cool carcasses quickly because salmonella bacteria can multiply at temperatures of about 7 °C or above.

Also, faster chilling can result in faster dispatch, less moisture loss and a longer shelf life.

Reducing the relative humidity in the chillers can reduce the likelihood of bacterial spoilage.

Muscle contraction can be reduced by keeping the muscle under tension during cooling. In the Tenderstretch process, hanging the carcasses by the aitchbone of the butt rather than by the Achilles tendon of the back leg keeps some of the more valuable muscles under tension during cooling, resulting in more tender meat in most of the more valuable parts of the carcass. This practice has to be applied within one and a half hours of slaughter and maintained for 20 hours. However, it has been little used, because it requires double handling at abattoirs, reduces the throughput from the cool rooms, and makes it more difficult for butchers to identify the individual muscles in the carcass. Also, any benefit has generally been confined to the carcasses from animals under four teeth.

Another practice called the Tender-cut method is to hang the carcass by the Achilles tendon and cut through the backbone at two places. This transfers the tension from the backbone to the meat.

The toughness of the meat from the loin can also be tougher if the carcasses are cut up earlier than two days after slaughter. In Denmark, when beef carcasses were broken down at various intervals after slaughter and the meat was chilled so that all the tests were done at the same time, the shear values for loin muscle were 21.5 kg for removal on day 1 and 10 to 11 kg for removal on days 2 to 6.

Passing an electric current through the carcass just after slaughter can prevent or reduce cold shortening and result in more tender meat. It results in an acidity level being reached in the muscle within one or two hours that would otherwise take 16 to 20 hours to develop, and the likelihood of cold shortening generally ceases when the pH level falls to about 6.3. It also probably shatters some chemical bonds within the muscles and may result in an early boost to the enzyme activity that tenderizes the meat during ageing.

The effect of electrical stimulation is generally greater in leaner carcasses, where cold shortening is more likely. Also, the degree of effect on meat tenderness varies between different muscles, with the main benefit generally being in the muscles of the hindquarters and along the back.

In American results, electrical stimulation followed by ageing improved the tenderness score for loin steaks by 44% compared to steaks that were frozen at two days of age and then cooked. The improvement with ageing alone was 26%.

Electrical stimulation can be done using a high- or low-voltage current. Use of the high-voltage current can also result in some improvement in the colour and texture of the meat during display and an increase in the shelf life by about half a day, but its use requires more care at the abattoir.

If electrical stimulation is excessive for the particular carcass, a condition known as heat shortening occurs, again adversely affecting the tenderness of the meat.

The proportion and strength of the connective tissue vary between muscles, and they are generally greater in hard-working muscles such as those of the legs. This, together with the effect of different cooling rates, affects the relative toughness of the meat from different parts of the carcass.

The meat is usually very tender in veal calves, changes little in tenderness in animals of about 9 to 48 months of age, and then slowly starts to become tougher and more variable in tenderness between individual animals beyond this age. In one set of results, the average shear values were 3.4, 4.8, 4.5, 5, 5.2, and 6.1 kg for cattle of 2, 9, 16, 27, 42, and 120 months old respectively. One researcher claimed that consistently tender beef cannot be assured from cattle with more than six permanent incisor teeth. A factor in this effect is chemical changes in the connective tissue making it tougher. Moreover, in very old cattle the muscle mass can decrease, resulting in an increase in the proportion of connective tissue.

In muscles that have not cold-shortened, the meat can be tenderized by ageing at a temperature just above the freezing point. Calpain enzymes in the meat weaken the chemical bonds holding the muscles in tension. Another enzyme, calpastatin, can retard the actions of the calpains, and a relatively high level of calpastatin appears to contribute to the generally tougher meat from tropical breeds of beef cattle. In one instance, the shear force value increased from 4.44 to 6.68 kg with an increase in the proportion of Brahman breeding in the steers from 0 to 75%. If the animals are no more than 50% of a tropical breed the meat is usually of acceptable tenderness.

The composition and tenderness of the meat can differ between apparently similar animals for reasons that have not been identified. The degree of inherent meat tenderness appears to be moderately heritable, and one claimed heritability value for overall meat tenderness was 0.6, indicating considerable genetic opportunity for selecting animals for more tender meat if it is practical to do so.

The effect of ageing is at a declining rate, and the results of experiments indicate that most of the possible benefit is likely to be obtained by the end of about two weeks if the holding temperature is about 3 °C. A higher temperature can result in faster ageing but may also result in spoilage and off flavours. In one set of results, the shear values were 4.3, 3.5, 3,1, 2.9, and 2.6 kg after ageing for 1, 7, 14, 21, and 36 days respectively. Meat that is cold-shortened will respond little to ageing, as is also the case with parts of the carcass with a relatively high content of connective tissue, including the forequarters.

Beef carcasses are now commonly broken down into joints within a few days of slaughter, and the joints are aged in plastic bags from which the air has been removed. This is the cryovac method. During this process, about 1–5% of the weight can be lost as moisture weep. Also, meat can be stored for up to about 12 weeks with this method, but with longer ageing off flavours can develop.

In contrast to the benefit of a longer ageing period found in experiments, when a large survey was conducted in several American cities on the tenderness of strip loin and sirloin steaks being retailed, there was no clear improvement in tenderness from ageing for more than seven days when the carcasses were broken down into joints one to three days after slaughter and the joints were aged for periods of different length. It was suggested that the lack of clear response to a longer period of ageing in the commercial situation may have resulted from higher temperatures during distribution and retailing accelerating the ageing process. Another finding was that the meat was generally tenderer from carcass grades involving greater fatness.

The enzymes involved in muscle formation in the live animal are also involved in tenderizing the meat during ageing. Therefore, there has been interest in whether faster weight gain just before slaughter would affect the enzyme activity and degree of tenderization of the meat after slaughter. The results from two American experiments indicated that faster weight gain just before slaughter is likely to improve the meat tenderness only if it continues long enough to appreciably affect the fatness of the carcass.

Aside from affecting the cooling rate of the carcass, it has been suggested that greater fatness can affect the tenderness by lubricating chewing and by reducing the concentration of connective tissue because it expands the muscle cells. In a review of numerous research papers, a clear overall trend was an improvement in all the eating properties of the meat with an increase in the thickness of fat cover on the carcass up to about 8 to 12 mm and with no pronounced further effect beyond this level.

Cooking the meat by wet methods, such as stewing, softens the connective tissue and results in tender meat. With dry cooking, such as grilling, the tenderness can be better if the meat is cooked rapidly and not overdone.

In some markets there is a preference for carcasses fattened to the point where pronounced amounts of fat appear through the meat, a condition called marbling. However, any benefit to the eating properties has been small. It can reduce the drying out of the meat during fierce cooking and help to lubricate chewing, as well as the carcasses fattened to this stage being unlikely to experience cold shortening. In one set of results, marbling improved the scores on average by 5% for tenderness and 16% for juiciness. Another set of results indicated that 50 days of intensive feeding after the animals had been reared at pasture would generally result in carcasses of favourable eating quality.

As described in an earlier chapter, Angus cattle are favoured in Australia for producing marbled beef, and in Japan the Wagyu breed has been especially selected for its marbling properties.

American results have indicated that the ease with which cattle develop marbling in a feedlot can be affected by such factors as the use of growth promotants and the rate of weight gain before entering the feedlot.

FLAVOUR AND JUICINESS

The flavour of the meat is blander from younger cattle than from older cattle and from grain-fed cattle compared to grazed cattle. Also, in America it has been found that there appears to be a preference for meat from cattle fed maize rather than other grains. Increased fatness also adds to the flavour of the meat. Moreover, the flavour of the meat can be affected by how the method and degree of cooking affect the production of breakdown products. Feeding cattle on lucerne pellets has also adversely affected the flavour.

Most of the differences in flavour of the meat from different species of animals are in the fat. Aside from having its own flavour characteristics, fat can be a source of absorbed taints. Untrained people eating very lean meat can sometimes have difficulty in identifying which species of animal the meat came from.

The juiciness of meat is affected by its fat and moisture content. Searing the outside of the meat in the early stages of grilling can help to retain moisture, and there can be less loss of moisture if the meat is cooked to the desired internal temperature more quickly. A greater degree of cooking can result in more moisture loss. Part of the contribution of increased fatness to juiciness seems to be that it stimulates the flow of saliva during chewing.

Longer storage of frozen or unfrozen meat can result in greater moisture loss and reduced juiciness.

Miscellaneous Factors

Bruises have to be removed from carcasses and the degree of bruising affects the value of the carcass. Bruising can be caused by various factors, including protruding objects, rough handling, and the presence of cattle with horns. In one set of results there was little bruising from saleyard selling, and in a second set most of the bruising occurred at the abattoir.

The livers from cattle affected with liver fluke can be condemned at abattoirs, and feeding hay to cattle too close to slaughter has resulted in tongue contamination with regurgitated rumen contents.

For manufacturing purposes, the meat from lean cows or bulls is preferred because it can be mixed with surplus fat from other sources. For minced beef products, such as hamburgers, a fat content of about 30% can result in optimum palatability in the cooked product.

Growth Hormone Implants

Growth hormone implants can be based on female or male hormones or a combination of each. They can increase the growth rate in proportion to the level of nutrition that applies and can be more effective in cattle after weaning. They can also sometimes have a slight effect on the fatness and eating properties of the meat, depending on the hormones used. Female hormones generally result in greater fatness and male hormones in reduced fatness. The product used should be varied according to what is being attempted.

It is vital to observe the stipulated withholding period for any cattle treated with chemicals or medicines and to declare the use of growth hormones. The use of growth hormones limits where the meat can be sold.

Consumer Response

Many consumers may not be able to differentiate precisely between beef differing in inherent quality.

In one American study, consumers could not differentiate between steaks of different shear values. Also, in another study, 71% of the participants preferred the steaks with a lower shear value and 22% had the opposite preference, with the acceptability of the steak of lower shear value being greater with a lower degree of cooking. Moreover, there was no relationship between the juiciness and flavour as assessed by the participants and those of a trained taste panel.

In this last survey, only 17% of those interviewed said they would be willing to pay more for more-tender steaks. This is not surprising because there are plenty of opportunities to obtain more-tender steaks by buying better cuts. It is more expensive to produce steak of more consistent eating quality because of the cost of the additional processes involved, but improved tenderness is likely to improve the relative competitiveness of beef against other meats.

MLA Programmes

The following programmes have been developed by MLA (Meat and Livestock Authority) in Australia to improve the marketing of beef:

1. *LPA (Livestock Production Assurance) programme.* In this programme, farmers have to declare any use of growth hormone implants and that stipulated withholding periods have been observed for any specified chemicals used. The name and address of the owner of the cattle has to be included and supported by numbered ear tags, which enable the cattle to be traced if necessary.
2. *AUS-MEAT programme.* This defines a method of describing carcasses to be used at accredited abattoirs. Values are included for animal maturity, method of feeding (pasture or grain), carcass weight, thickness of fat cover, meat colour, degree of

marbling, and the degree of carcass trimming. Measurement of fat colour and eye muscle area can also be provided if requested. Moreover, the programme provides display posters of meat cuts and the proportion of each cut for a sample carcass.
3. *MSA (Meat Standards Australia) programme.* This programme is based on extensive research and is applied at accredited abattoirs by trained assessors. The aim is to estimate the eventual eating qualities of different cuts of meat from the carcass, using measurements of carcass weight, sex, meat and fat colour, proportion of tropical breed according to the size of any hump present, hanging method (Achilles heel or tenderstretch), use of hormone implants, ossification, meat pH, and temperature. For the carcasses to be acceptable, the pH must be 5.71 or lower, and to obtain accurate readings the meat temperature must be no higher than 12 °C.

After these readings have been taken the carcass is broken down into cuts which are stored in cryovac bags. A quality score and suitable cooking method are predicted for each cut after 5 or 14 days of ageing, with a better score commonly applying for the longer ageing period. These scores are attached to the bags of meat sent to retailers.

The degree of ossification rather than teeth numbers is used because it is a more reliable indicator of the level of maturity. It is assessed at three places along the backbone as the degree to which cartilage present in young animals has been replaced by bone in more-mature animals. More-mature animals can have tougher meat because of chemical changes in the connective tissue of the muscles. Rearing on prolonged poor nutrition can result in a greater maturity score relative to carcass weight. Also, patches of poor fat cover can result in localized cold shortening.

Conclusions

A farmer's role in producing good-quality beef is to produce carcasses of the weight, fatness, and animal age preferred by buyers and to do so by using quiet cattle without horns. Aside from that, cattle should be fed and handled appropriately when being sent to market, and the designated withholding periods should be observed for the use of various chemicals and medicines, together with a declaration of any use of growth promotants.

CHAPTER 17

Parasites

Parasites can affect cattle performance to varying degrees in different locations. The important parasites in cattle in southern Australia are roundworms, liver fluke, and lice.

ROUNDWORMS

The severity of the worm problem is generally in direct proportion to the annual rainfall, both between districts and between years within each district. In situations of moderate rainfall, worms are generally of little concern.

The common symptoms of a major worm infestation are ill thrift, a dry coat, watery dung, and sometimes a change in coat colour. The degree of effect on the animal varies according to the size of the worm burden.

The most important worm species in South Australia *Ostertagia ostertagi* (the small brown stomach worm) and *Trichostrongylus axei*. Species of generally lesser importance are *Haemonchus placei*, *Cooperia punctata*, *Bunostomum phlebotomum* (the hookworm), and *Oesophagostomum radiatum* (the nodular worm) in regions of summer rainfall and *Trichostrongylus axei* and *Cooperia oncophora* in regions of winter rainfall.

Cattle can also be infected with lungworms and stomach worms. Lungworms are generally unimportant in Australia, and stomach worms are found mainly in coastal regions.

Eggs produced by the adult worms in the digestive tract of the animal pass out in the dung, and if the temperature and moisture conditions in the pasture are favourable these eggs may develop immediately into infective larvae. These infective larvae move up the pasture plants on the surface moisture present and are consumed by the animals during grazing. Worm eggs can survive in dry dung pads for up to about a year under some situations, and worm larvae can survive in the pasture for a few months under suitable moisture and temperature levels.

In regions of winter rainfall there can be rapid development of worm larvae during autumn and spring, and slow development of some species during winter if the weather is not too cold. Larvae which develop in autumn can survive through to spring. During summer there can be rapid development of larvae in summer rainfall regions, but no development in regions normally dry at this time even after rain.

In the animal, the ingested larvae may attach to the gut wall and start to develop immediately into mature worms and reach an egg-producing capacity after about three weeks. Alternatively, the larvae may become embedded in the gut wall in an inhibited state for periods of varying duration. Inhibited larvae can be especially important with the *Ostertagia* species of worm, where large numbers of larvae can become embedded in the gut wall from late winter to early summer in winter rainfall regions and remain there until stimulated to develop and emerge over a few days in autumn. In the dormant state these worm larvae have been difficult to treat in the past, and if large numbers suddenly emerge in autumn there can be considerable damage inflicted on the gut wall. Also, the emergence of these larvae in autumn, together with the large number of infective larvae that can be emerging from dung pats, can result in massive worm burdens in cattle at this time of year.

What causes worm larvae to become inhibited and later emerge is not clear. Modern anthelmintics are more effective against inhibited larvae.

Cattle generally become more resistant to worms as they become older, unlike what generally applies in sheep and goats. In cattle, a substantial resistance has generally developed by about 20 to 24 months of age, and better nutrition can hasten this development as well as directly benefiting the animal's capacity to withstand the effects of worms. However, poor nutrition can weaken the resistance of mature cattle to worms, and bulls can be relatively susceptible to worm infestation. Moreover, older cattle that have had little exposure to worms when in a dry region can be highly susceptible to infestation if moved to a wetter region.

Use of a heavier stocking rate could be expected to increase worm burdens in grazing animals by resulting in a higher degree of pasture contamination, poorer nutrition, and greater inducement to graze closer to dung pats. On the other hand, with a heavier stocking rate the worm larvae are more exposed to weather effects.

With sheep, the effect of stocking rate on the worm burdens has been important for *Nematodirus* species of worm, slight for *Haemonchus* species, and unimportant for the others. With cattle, there has been a slight increase in the presence of *Ostertagia* species of worm with an increase in stocking rate.

In two American experiments, rotational grazing combined with a heavier stocking rate resulted in heavier worm burdens in cattle, presumably because the cattle were induced to graze closer to dung pats before being moved. Also, in Britain pasture-harrowing resulted in heavier worm burdens in cattle, presumably because of reduced opportunities to avoid patches of high larval infestation.

Worm eggs can persist in a pasture for up to about a year, and this affects the merit of short-term spelling. It has been claimed that the length of spelling period required to obtain moderately clean pastures is 4–5 months during autumn and winter or 3–4 months during the rest of the growing season. Dung pats persist longer in drier regions, and this may affect the length of spelling period desirable.

Some species of worm do not transmit readily between sheep and cattle, and this effect can be used in helping to control worm problems. There is virtually no cross infestation between sheep and cattle for the *Ostertagia, Oesophagostomum, Nematodirus, and Bunostomum* species of worm but ready cross infestation with the *Trichostrongylus axei* and *Haemonchus contortus* and *Haemonchus placei* worms.

Cattle are treated against worms by administering an anthelmintic as an oral drench, injection, or pour-on product.

There are three main groups of anthelmintics currently being used, and within each group there are different formulations. These groups are the following:

a) levamisole, which is a clear product available as an oral drench or pour-on product
b) benzimidazole, which is a white product that is mainly available as an oral drench but can also be obtained in a pour-on formulation
c) macrocyclic lactone (mectin), which is available in an injectable or pour-on formulation.

Of the three anthelmintic groups, the levamisole is generally poorly effective against immature worms.

Worms can slowly develop a resistance to a particular anthelmintic, but this effect has not been evident in cattle, unlike what has applied with sheep. Ways to reduce the likelihood of a resistance developing are to rotate between different anthelmintics and administer them carefully so that a full dose enters the animal's body.

Worm egg counts conducted at a laboratory on samples of cattle dung can provide a reasonable estimate in most instances of the worm burdens in the animals, but in a few instances cattle with heavy worm burdens can show low egg counts.

A general recommendation is that only young cattle should be treated against worms except where mature cattle show evidence of substantial worm burdens.

Despite the recorded resistance in mature cattle to worm infestation, in the region of western Victoria with relatively high rainfall, it has been found advisable to treat first and second calvers. Also, responses have been recorded in America in mature cows of good to excellent body condition. In the different studies, the degree of responses varied from 2.6 to 12% for the pregnancy level, from 5 to 17kg for the weaning weight and in one study there was a 27% increase in milk production. Moreover, with mature cows set-stocked on annual pasture in experiments at Rutherglen the texture of their dung indicated both a need for treatment and a response to it.

A recommendation for winter rainfall regions is to treat susceptible cattle shortly after the opening rains and move them to a relatively clean paddock. Paddocks in order of decreasing cleanliness are likely to be those not grazed for a year, those grazed for at least the previous four months by sheep, and those grazed by mature cattle. The most heavily contaminated paddocks are likely to be those that have just been grazed by young cattle.

To avoid carrying over contamination, cattle should not be moved to a clean paddock until one or two days after treatment.

Where only set-stocking can be practised, the treatment should as far as is practical be at a time of year when it is likely to be most effective. Treatment as soon as the pasture dries off will be followed by several months of no further contamination and no pickup or larvae, and treatment in spring after the pasture starts to grow rapidly is likely to result in reduced consumption of larvae because of less inducement to graze close to dung pats.

Because of the differences in the environment between different locations, it is advisable to obtain local advice on favourable treatment practices.

Liver Fluke

Liver fluke can be a serious problem in cattle with access to grazing on permanently damp patches.

Liver fluke are distinctively shaped organisms which are found mainly in the ducts of the liver but can also be found in some other organs. Infestation can be evident as a dry coat, a swelling under the jaw, ill thrift, and lighter-coloured coats in Herefords and some other breeds. Also, infected livers can be condemned when the animals are slaughtered, and an invasion of the damaged bile ducts by clostridium bacteria can cause black disease, leading to sudden death.

Eggs produced by the adult fluke pass out of the animal, hatch within about ten days at a temperature of 10 °C or above to form miracidia, which must within a few hours find and enter a host snail of the species *Lymnaea tomentosa*. Within the snail, a miracidium develops and multiplies to produce rediae, which pass out of the snail, swim until they can attach to plants, and become encysted. If the infected pasture is eaten, cysts enter the animals, where they may remain dormant in the liver for up to about ten weeks before entering the bile ducts and developing into adult fluke.

The snails involved in liver fluke transmission are about 5 to 25 mm long, and when placed on a flat surface with the shell opening downwards and viewed from above the whorl widens in a clockwise direction.

There is a range of products available for treating liver fluke. Some of these products are effective against both immature and adult fluke, and some are effective only against adult fluke. Products are now also available which combine worm and fluke treatments. Draining of wet patches can be beneficial, but applying chemicals to kill the snails has proved to be generally ineffective.

As is the case with worms, cattle generally develop a substantial degree of resistance to liver fluke with prolonged exposure.

The best treatment times seems to vary between the different environments, so local advice should be sought on the recommended treatment practices.

Cattle can be protected against the black disease associated with the presence of liver fluke by vaccination. If a 5-in-1 or 7-in-1 anticlostridium vaccine is used regularly this can also prevent the odd cattle death that can occur periodically from a cause that can be difficult to diagnose.

Lice

Heavy lice burdens are commonly found in cattle in winter, especially in high-to-medium rainfall zones. This causes the animals to persistently rub themselves against solid objects.

There are biting and sucking lice, with a total of six species in Australia, but not all these species are present at any one location. The biting lice live on the skin, and in large numbers can cause anaemia. There is debate about how badly lice affect animal performance, and in some instances there appears to have been little or no effect on the weight gain. However, important considerations are whether a farmer wants to see his animals experiencing constant irritation, and the possible effects on hide damage and poorer appearance of the animals if they are to be sold. Cattle rubbing themselves on trees can sometimes also cause ring-barking.

The different species of lice can vary in length from about 1.5 to about 5 mm long. The eggs take about 8 to 19 days to hatch, and maturity is reached about 3 to 6 weeks later.

The current methods of treatment are to treat with an insecticide that is sprayed on, applied as a pour-on product, or impregnated in ear tags. The cheapest treatment, excluding handling costs, is sprayed-on insecticide. Spraying can be the most effective treatment if used properly. Spraying with fine mist can result in effective wetting of the animal with economical use of the chemical. With spraying, two treatments about three weeks apart can kill the lice that have emerged after the first treatment.

Conclusions

Worms can be a greater problem with higher rainfall, and liver fluke live in wet spots. The most sensitive animals to infestation are bulls, younger and poorly fed cattle, and those just brought in from a drier region.

In pest control it is important to use effective products, apply them at appropriate times relative to the seasonal rainfall pattern, dose carefully, rotate between different chemicals, and especially in the case of worm control return the animals to clean paddocks.

Lice infestation affects the comfort and appearance of the animals but may not affect the weight gain.

CHAPTER 18

Drought Feeding

Drought occurs with different frequencies and different durations in different regions, and the long-term weather records for the region can indicate what may occur. There is no certainty of what will happen, and decisions have to be made on the basis of probability according to the best information available.

Feeding Practices

The drought feeding practice in a region is likely to vary according to how long the drought may last. If a drought usually lasts for only a year then the aim may be to feed so that about normal production will resume as soon as the drought breaks. On the other hand, if droughts sometimes last longer than a year then the aim may be to preserve the most valuable breeding stock or to use feed reserves to buy fresh stock just before the drought is expected to break.

Only the animals that are likely to be profitable to retain should be fed during a drought. Earlier selling is likely to result in better prices. In situations of uncertainty a sound practice can be to sell progressively until the situation becomes clearer.

There is usually little consistency in the weather results between years but close examination of the long-term results can sometimes reveal a very valuable piece of information. For example, as described in another chapter, in north-eastern Victoria little rain by the end of June can be an almost certain predictor of a dry year.

Unless prime stock are nearly fat to begin with it can be unprofitable to fatten cattle during a drought in anticipation of higher beef prices, because at least in some parts of Australia beef prices often rise only moderately at that time. This probably occurs because of the number of cattle being sold and resistance by the general public to paying higher meat prices. Thus, rather than putting feed into fattening animal it can sometimes be more profitable to sell these animals as well as the feed they would otherwise be given. Also, in some regions at least, the prices for store stock often rise only moderately at the end of a drought, probably because of a shortage of funds to buy them and farmers temporarily changing to other enterprises. However, there is no certainty about what will actually occur. If in doubt, a sound practice can be to select a bit of each option.

The most common drought feeds for cattle are hay, silage, and grain. Grain is a less convenient and less safe feed than hay or silage, but relative to its feeding value it can be the cheapest feed. Sometimes various by-products are available on a limited scale.

When just used to maintain weight the relative values of different feeds are generally proportional to their ME (metabolizable energy value) measured as megajoules per kilogram (MJ/kg). Some feeds, such as straw and low-quality pasture, may be of inadequate nutritional value to achieve even animal maintenance. Pea straw can have about as much as three times the crude protein level of cereal straw but a digestibility similar only to oaten straw.

Information on the nutritional value of various feeds can currently be found on the Internet in the Primefact bulletin published by the New South Wales Department of Primary Industries. The daily rations (kg/head) suggested for different classes of cattle when the feeds are average quality grain, hay, or silage are as follows:

	Grain (12 ME)*	Hay (8.5 ME)	Silage (9 ME, 30% DM)
Weaners (200 kg)	2.5	3.5	12
Yearlings (250 kg)	3	4	15
Dry adults (400 kg)	4	6	20
Late pregnancy (425 kg)	5	8.5	27
Lactating (425 kg)	-	10.5	30

*ME/kg

With these rations it is also recommended that they be increased by 20% during cold weather, using hay if possible, and that if grain is to be fed to lactating cows the rations should be at least 1.5 kg of hay as well as 6.5 kg of grain. This is because cows can milk poorly and have a relatively low butterfat content in their milk if fed little roughage. Also, it is desirable that the ration should contain at least 9% of crude protein for young stock and 11% for lactating cows.

At low levels of feeding, oats can have about the same nutritional value as wheat and barley, but at higher levels of feeding the nutritional value value of oats is substantially less than that of these other two grains.

The nutritional values of individual feeds from different sources vary substantially, and the only way to know the actual value is to have the feed tested at one of the laboratories providing this service.

In one experiment, the digestibility of barley straw was improved from 50.4 to 56.5% and the daily amount consumed by the sheep increased from 346 to 470 g by using a pump and steel pipe to inject the straw with urea at 2.2% and molasses at 0.6% of the weight of the straw. This solution must be injected evenly throughout the straw to avoid urea toxicity.

If the feeding is to be in troughs it is likely to be profitable to crush or roll wheat and barley but not oats. More digestive upset is likely with crushed or rolled grain, and where the grain is to be processed it should only be coarsely cracked. Also, grain is low in calcium, so 1.5% of limestone should be added to the grain. With whole grain, the limestone is likely to stick to the grain better if the grain is slightly moistened.

In intensive feeding, 1% of salt is usually added to the grain as well as limestone, but Queensland results indicated that the addition of salt to the grain in drought rations can lead to oedema.

To reduce the amount of wastage it is best to provide the grain in improvised troughs. If the herd is not too large, a temporary holding yard and feeding yard used in combination can reduce the amount of trough space required. Also, it is important to provide enough trough space so that all the animals receive a fair share of the feed.

If the grain is to be fed on the ground, the amount of wastage can be reduced if the grain is laid out in heaps of about 20 kg rather than spread as a trail.

Cattle need to be conditioned to grain digestion over at least two weeks after all the cattle have started to eat grain. Placing hay on top of the grain can reduce the time taken to get all the cattle to start eating grain. During conditioning the amount of hay should be reduced progressively and the amount of grain increased. After the cattle are well conditioned to grain digestion the frequency of feeding can be changed progressively to twice a week if desired. In one study, while feeding twice a week was considered satisfactory, the death rate was higher than with daily feeding.

Many farmers seem to have been successful in using grain-only rations for drought feeding whereas others have had problems with these rations. Also, in an experiment in Rutherglen with weaner steers fed individually on daily rations of whole oats, whole wheat, or rolled wheat at levels near the maintenance requirement after four weeks of conditioning, the animals could not be kept healthy on all-grain rations. The inclusion of 0.5 kg of hay in each daily ration immediately remedied this problem. Moreover, in an experiment conducted elsewhere, calves fed only grain had relatively poor weight gains and suffered from bloat. On the other hand, at Hamilton, providing hay in addition to grain twice a week was of no great benefit in improving the generally unsatisfactory situation that applied.

This difference in result between Rutherglen and Hamilton may have been due to a difference in when the hay was provided. Rumination can cease when only grain is provided, and this affects feed movement in the rumen. Also, the chewing of hay or other long roughage stimulates the flow of saliva, and saliva contains phosphates and bicarbonates that help to moderate the high levels of rumen acidity that occurs with grain-feeding. Therefore, the inclusion of roughage in the diet is likely to be much more effective if it is provided shortly after the grain has been eaten.

In Britain, as described earlier, when calves were fed ground and pelleted dried grass they could not be kept healthy unless some chopped dried grass was also included in their diet to stimulate rumination.

At Hamilton, when calves of 8–10 months old were fed drought rations of whole wheat, crushed wheat, or whole oats, the average daily weight loss did not differ between treatments but tended to be greatest with the crushed wheat and least with the whole oats. The calves would not eat all their rations, and there was evidence of acidosis with crushed wheat. In a second experiment, the total weight change was better on crushed wheat than on whole wheat when hay was included in the diet.

Reports with sheep and cattle indicate that where some paddock roughage remains, all-grain rations can be used quite satisfactorily.

When cattle are on all-grain rations it is vital not to change from one grain sample to another quickly, and to ensure that water is available continuously. Also, cattle on all-grain rations can be highly susceptible to the effects of cold, wet weather, and the New South Wales Department of Primary Industries has advised that some hay should be provided when such weather is imminent. On a farm near Albury, a substantial number of cows being fed grain-only drought rations died during an unseasonable spell of cold, wet weather in summer. Moreover, vets have warned that at the end of a drought it can be unsafe to end hand-feeding before the cattle have had time to re-adapt to grazing and digesting pasture.

If desired, calves can be weaned at 6 weeks of age but preferably not before 8 weeks. After weaning, they should be given calf pellets and good-quality hay. Also, providing suckling calves with creep feeding from about 4 weeks of age onwards during a drought is likely to be an efficient practice, but care is needed to prevent acidosis.

When non-lactating cattle are on drought rations their feed requirement to maintain weight can be about 20% lower than normal, probably because of shrinkage in the size of the digestive tract and liver, two organs with high energy expenditure.

The availability of water can be a great problem during drought. Where the water is provided from dams an electric fence can be used to limit access to the dam and thereby prevent the fouling of the water that remains. With dams there is also a risk of cattle becoming dangerously bogged when the water level is low. Carting water to cattle during a drought can be extremely demanding.

The water requirement of cattle can be about 10 l a day per 100 kg and even more during very hot weather.

In recommendations published on the Internet, the New South Wales Department of Primary Industries advises that at the start of drought feeding all the cattle should be treated for internal parasites, vaccinated against clostridial diseases, and given an injection of vitamins A, D, and E. It also recommends that early-weaned calves should be vaccinated against coccidiosis. However, any access to a green pick or green feed at some time will reduce the need for a vitamin injection.

Drought conditions can sometimes induce animals to eat toxic plants that would normally be avoided, such as cypress hedges and bracken.

Conclusions

Overall, in drought management it is important to use long-term weather records to try to identify the likely conditions that may be encountered. It is also important to be selective in the stock retained and to adjust the feeding practices so that they are reasonably safe. Adverse effects can occur from lack of roughage, cold spells, quick feed changes, temporary lack of water, and stopping hand-feeding too early. Estimates of the likely effects in livestock markets during and at the end of a drought should be based on facts and not supposition.

APPENDIX

Scoring Methods of Beef Cow Condition

a) East of Scotland Agricultural College Method

Score	Description
0	Spine is very prominent, and ends of short ribs at the loin feel very sharp with no fat cover.
0	Spine is prominent, and ends of short ribs feel sharp with little fat cover.
1	Short ribs can be felt but are rounded with a thin covering of fat.
2	Individual short ribs can be felt only with firm pressure.
3	Short ribs cannot be felt.
4	Short ribs have a thick layer of fat cover.

b) American Method

Score	Description
1	Severely emaciated, physically weak, bones easily seen, mainly found in sick animals.

2	Severely emaciated but not very weak, bones easily seen, muscle tissue severely depleted through the hindquarters and shoulders.
3	Very thin, no fat on ribs and brisket, backbone easily seen, and some muscle depletion on hindquarters.
4	Appears thin, ribs easily visible, backbone showing, short ribs at loin are still very sharp and barely visible individually, and muscle tissue not depleted on the shoulders or hindquarters.
5	Moderate-to-thin body condition, last two ribs can be seen, little evidence of fat around brisket and over the ribs and at the tail head, short ribs are now smooth and no longer individually identifiable.
6	Good and smooth appearance, some fat on brisket and tail head, back appears rounded, and fat can be felt over the ribs and short ribs.
7	Very good condition, brisket full, pockets of fat at tail head, ribs smooth and soft to handle because of fat cover.
8	Obese, neck is thick and short, back is very square because of excessive fat, brisket is distended, and heavy fat pockets around tail head.
9	Extremely obese, heavy fat depots in udder.

ABOUT THE AUTHOR

The author graduated in agriculture and agricultural economics from Glasgow University and obtained a master of agricultural science degree from University of Melbourne.

After first graduating, he spent two and a half years as assistant to a farmer in England, followed by six years managing a group of farms in Western Australia, where the enterprises included stud and commercial beef cattle. After that, he spent 30 years as a beef cattle researcher at Rutherglen Research Institute in north-eastern Victoria and, on early retirement, share-farmed with beef cattle on a farm at Rutherglen.

Mr Hamilton can be contacted by telephone on 02-60329361 or by email on david.lauraine@bigpond.com.

www.ingramcontent.com/pod-product-compliance
Lightning Source LLC
Chambersburg PA
CBHW030941180526
45163CB00002B/657